3rd International Conference on Robotics and Mechatronics (ICRoM 2019)

IOP Conference Series: Materials Science and Engineering
Volume 637

Sabah, Malaysia
9 – 11 August 2019

ISBN: 978-1-5108-9839-4
ISSN: 1757-8981

Printed from e-media with permission by:

Curran Associates, Inc.
57 Morehouse Lane
Red Hook, NY 12571

proceedings.com

Some format issues inherent in the e-media version may also appear in this print version.

Institute of Physics
Dirac House, Temple Back
Bristol BS1 6BE UK

Phone: 44 1 17 929 7481
Fax: 44 1 17 920 0979

techtracking@iop.org

Additional copies of this publication are available from:

Curran Associates, Inc.
57 Morehouse Lane
Red Hook, NY 12571 USA
Phone: 845-758-0400
Fax: 845-758-2633
Email: curran@proceedings.com
Web: www.proceedings.com

Table of contents

Volume 637

3rd International Conference on Robotics and Mechatronics (ICRoM 2019)

9–11 August 2019, Sabah, Malaysia

Accepted papers received: 2 September 2019
Published online: 11 October 2019

Preface

Papers

Robotics

Mechanical Engineering

IOP Conf. Series: Materials Science and Engineering **637** (2019) 011001 doi:10.1088/1757-899X/637/1/011001

Preface

The 3rd International Conference on Robotics and Mechatronics (ICRoM 2019) was held on August 9-11, 2019, in the Horizon Hotel in the beautiful city of Kota Kinabalu Sabah. Since its inception, ICRoM has attracted many researchers from all over the world to share their findings. The delegates are from the USA, England, Czech Republic, Iraq, China, Malaysia, Thailand, and India.

Despite ICRoM 2019 aims for different topics, the conference has been successfully gathered different researchers to discuss ideas for multidiscipline research which are required for current and future technology. In addition, with the advancement of many aspects of engineering and technologies, we believe that ICRoM 2019 has played a vital role to contribute a significant knowledge and practical implementations particularly in Manufacturing Technology as well as Automation and Mechatronics Engineering.

We would like to thank all excellent Keynote Speakers and Invited Speakers which have delivered and shared their current research, respectively. The presentation started with Prof. Kok-Meng Lee (Georgia Institute of Technology, USA), followed by Prof. Li Lu (National University of Singapore, Singapore), Prof. Sujan Debnath (Head of Mechanical Engineering, Curtin University, Malaysia), Assoc. Prof. Vincent Lee Chieng Chen (Curtin University, Malaysia), Prof. Dr. Eric Dimla (RMIT University Vietnam, Vietnam), Assoc. Prof. Petr Valasek (Czech University of Life Science Prague, Czech Republic), Assoc. Prof. Jamaludin Jalani (Universiti Tun Hussein Onn Malaysia, Malaysia) and finally distinguished oral presentations.

We also would like to thank all the reviewers and proceeding team for their time and effort to ensure the proceeding document are in good quality. We sincerely hope that the proceeding will benefit to the people brought by the academic community and industrial practitioners. Continuous supports are essential from various parties particularly academicians, students, researchers, professionals, engineers, and scientists from academia and industry to ensure even more meaningful conferences are served in the future with numerous critical idea exchanges, diverse opportunities for fruitful networking and future collaborations between the delegates.

Warmest regards,

Conference Program Chair
Assoc. Prof. Jamaludin Jalani
Universiti Tun Hussein Onn Malaysia, Malaysia

IOP Conf. Series: Materials Science and Engineering **637** (2019) 011002 doi:10.1088/1757-899X/637/1/011002

Conference Committees

Conference Chair

Sujan Debnath, Curtin University, Sarawak, Malaysia

Conference Program Chairs

Erween Abd Rahim, Universiti Tun Hussein Onn Malaysia, Malaysia

Jamaludin Jalani, Universiti Tun Hussein Onn Malaysia, Malaysia

Vincent Lee Chieng Chen, Curtin University, Malaysia

Eric Dimla, RMIT University Vietnam, Vietnam

Local Organizing Committees

Mohd Suffian b Misaran, Universiti Malaysia Sabah, Malaysia

Choong Wai Heng, Universiti Malaysia Sabah, Malaysia

Mohd Adzrie Radzali, Universiti Malaysia Sabah, Malaysia

Mohd Amran hj Madlan, Universiti Malaysia Sabah, Malaysia

Technical Committees

Alokesh Pramanik, Curtin University, Australia

Lin Mingxing, Shandong University, China

Guohua Cui, Hebei University of Engineering, China

En-Chih Chang, I-Shou University, Taiwan

John M. Ikome, Vaal University of Technology, South Africa

Moola Mohan Reddy, Curtin University, Sarawak, Malaysia

Anika Zafiah Mohd Rus, Universiti Tun Hussien Onn Malaysia, Malaysia

Abu Bakar bin Ibrahim, Sultan Idris Education University, Malaysia

Lola Domnina Pestaño, University of Santo Tomas, Philippines

Yap Hwa Jen, University of Malaya, Malaysia

Baharudin B.T.H.T, Universiti Putra Malaysia, Malaysia

MP Jahan, Miami University, USA

Asma Perveen, Nazaryabev University, Kazakhistan

Anshuman Srivastava, Mechanical Engg Dept, SIET, Allahabad, India

IOP Conf. Series: Materials Science and Engineering **637** (2019) 011002 doi:10.1088/1757-899X/637/1/011002

Yang Xi, Newomics Inc., USA

M. Felix Orlando, Indian Institute of Technology Roorkee, India

Bidyadhar Subudhi, National Institute of Technology, India

IOP Conf. Series: Materials Science and Engineering **637** (2019) 011001 doi:10.1088/1757-899X/637/1/011001

Statement of Peer Review

All papers published in this volume of Materials Science and Engineering have been peer reviewed through processes administered by the proceedings Editors. Reviews were conducted by expert referees to the professional and scientific standards expected of a proceedings journal published by IOP Publishing.

Development of Two-Fingered Underactuated Robot Gripper using 3D Printer

Amirul Syafiq Sadun, Jamaludin Jalani, Siti Hana Nasir, Mohd Nazrul Roslan, Hairulazwan Hashim, Syazwani Mohd Anuar.

Faculty of Engineering Technology,
Universiti Tun Hussein Onn, Kampus Pagoh, Malaysia
amirul@uthm.edu.my, jamalj@uthm.edu.my, sitihana@uthm.edu.my,
nazrul@uthm.edu.my, azwan@uthm.edu.my, syazwani0501@gmail.com.

Abstract. This research present the development of a two-fingered underactuated robot gripper by using 3D printed machine with a material of Polylactic acid or PLA. The gripper design consist of two individual servo motors that drives the finger movement to perform a certain amount of grasping force. The new design of the finger using Solidworks was inspired by several existing robot gripper however with a low cost approach to obtain a good finger tracking and angular control. The fully fabricated and assembled gripper was then tested to determine the operation results during basic grasping and ungrasping by connected to Arduino Mega and MATLAB software. The angular data during the operation shows that the gripper successfully perform a stable operations despite some minor disturbance due to the mechanical defect on the servo plate. Further analysis will be conducted in the future in order to identify the full capability of the gripper in real world application such as force control operations.

Keywords: Robot Hand, Gripper, 3D printing, Grasping, Functional Needs Design

1.0 Introduction

Robotic application has been widely used in various industries to reduce labor cost by replacing human with robots in the production line since robot offer superior performance in terms of accuracy, precision, rigidity, and most importantly speeds. Robot arm are often used to pick and transport an object from one place to another. Since robot movement could cause fatigue, it may cause damage to the robot structure. Gripper is the main mechanism used to pick up an object. It is one of the incomplete parts from the robotic arm where it consists of shoulder, elbow and wrist. Grasping mechanism design has been the focus of several researchers in the field of robotics [1] [2] [3]. In addition to other existing limitation (such as backlash and frictions), a robot gripper grasp of an object are restricted within regular geometries [1]. Especially for 2 finger robot gripper where it has limited force that can be applied with limited shape to grasp. Designing gripper fingers properly can increase the work cell throughput, overcome robot inaccuracy and enhance overall system performance.

2.0 Background Review

In this section, useful components for this projects such as the robotic gripper mechanism, sensors and actuators involved are explained in detail.

2.1 Underactuated Gripper Mechanism

In general, there are two types of robot gripper design which are fully actuated and underactuated as shown in figure 1 [4] [5]. The fully actuated designs tend to shapes like human hand (humanoid) where the underactuated design tend to look more industrial application [1].

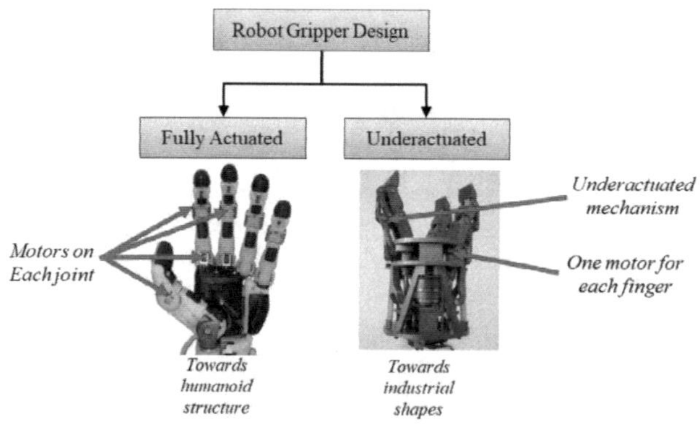

Figure 1. Robot Gripper Design

For underactuated designs, a two finger grippers could perform optimally for a single specific task with small rectangular or cylindrical objects. However, it worth noting that this type of gripper has a geometrical limitations for grasping an object with a complex shapes [2]. Due to these limitations, researchers make an effort dedicated to the development of multiple finger underactuated robot hands/grippers. These devices tries to mimic one or more characteristics of the human hand [6]. The three-finger gripper has existed for some time, but only been widely used recently. In the future, robots are more flexible and will be able to handle wider range of tasks without changing its parts. The most common used gripper would be two finger gripper as it is easy to handle and operate and has a better grasping for parallel gripper.

2.2 Servo Motors

This is the most commonly used in actuators for the robot hand/gripper prototype. The motor is light weight and equipped with DC motor and gear system which allows it to produce a decent amount of torque (depending on the motor pricing). The servo motors are shown in figure 2.

Figure 2. The DC Servo Motors

IOP Conf. Series: Materials Science and Engineering **637** (2019) 012001 doi:10.1088/1757-899X/637/1/012001

3.0 Design and Fabrication

This section describes the overall method of the project including the robot design, setup testing and analysis. The design and development stage for the Robot gripper is shown in figure 3.

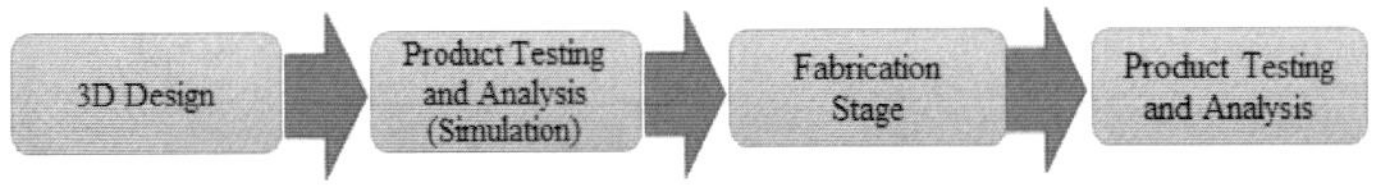

Figure 3. The Design Process

The 3D design and components of the gripper are illustrated in figure 4. The drawing consist of main component such as finger links and servo motor placement. The universal attachment lock was also considered in the design as this robot can be attached to any other design parts such as robot arm in the future [7].

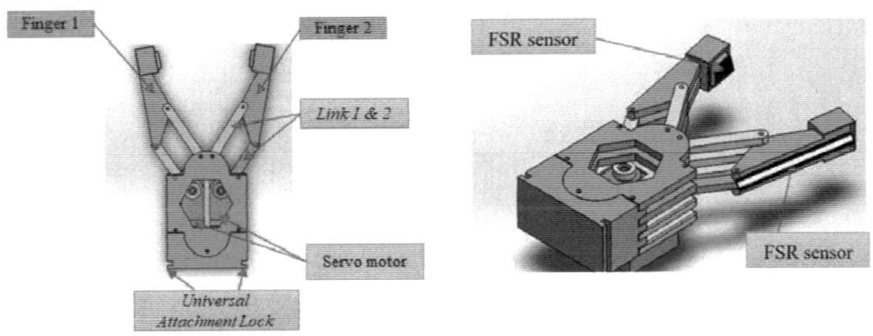

Figure 4. The *Robot gripper 3D Design*

These main components that are important and directly affect the physical operation and properties of the grippers. The design also considered the potential force sensor placement on the gripper for the next stage analysis, however the results of the testing are not discussed in this paper. The fabrication process is shown in figure 5 where the completion of all parts took approximately 16 hours (using the Cura Ultimaker software). The material used was the Polylactic acid or PLA which is common material for prototyping.

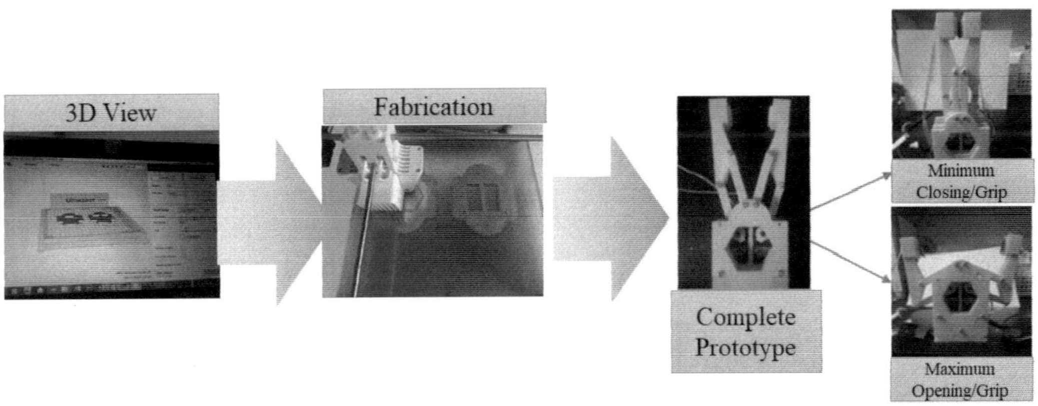

Figure 5. The *Fabrication Stage*

IOP Conf. Series: Materials Science and Engineering **637** (2019) 012001 doi:10.1088/1757-899X/637/1/012001

4.0 Prototype Testing and Analysis

The hardware setup for the testing and analysis are shown in figure 6. Similar setup can be seen in [8].

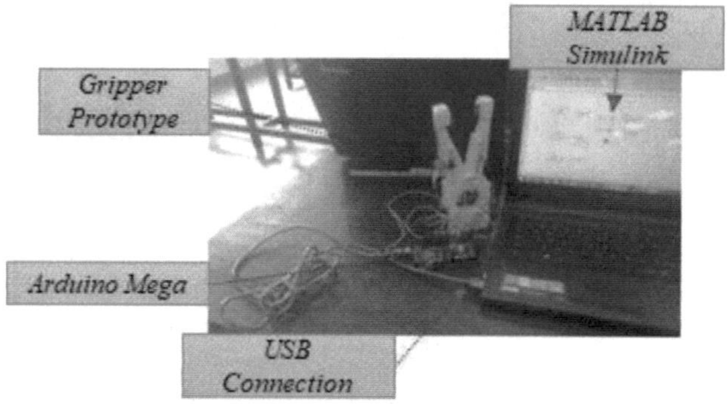

Figure 6. The Hardware Setup for Testing and Analysis

Based on the test results, it was found that the gripper capable of moving within the θ1 and θ2 range of 25 degrees until 180 degrees as shown in figure 7. The θ1 and θ2 also represents the maximum and minimum opening respectively.

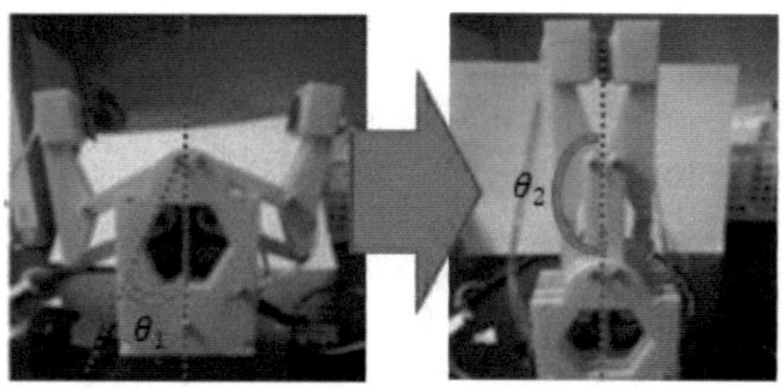

Figure 7. The Actual Closing and Opening Angle (θ1 and θ2)

Moreover, the operation results for repeatable task between θ1 and θ2 (25° to 180°) for 30 seconds are shown in figure 8 below.

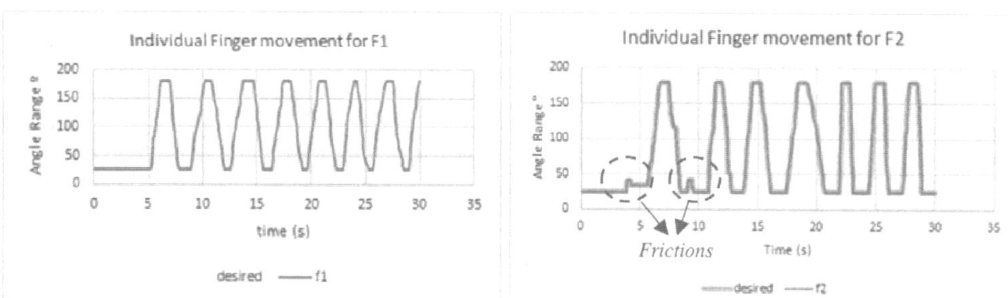

Figure 8. Individual Finger Movement Finger 1 (f1) and Finger 2 (f2)

Based on the results, there is not much difference for the angle range between desired position and the actual gripper position. However, the results at 4s and 8s shows the finger does not return to 25° due to the mechanical defect on the servo plate that hold the servo motor from returning the to the original state. There are also minor fluctuation detected on Finger 2 (f2) due to frictions on the links.

5.0 Conclusion and Recommendation

The overall process which involve the product design, fabrication, testing and analysis has been successfully conducted in the development stage of the prototype gripper. The proposed new design has shown a promising results whereby it capable of performing grasping and ungrasping operation smoothly. The gripper maximum opening has been verified to the extent of $\theta 1 = 25^{o}$ while the minimum opening is $\theta 2 = 180^{o}$. The simple underactuated mechanism of the prototype were able to provide a decent grasping force for grasping operation. However, the topic of grasping force will not be discuss further in this paper. The design of universal attachment lock has also been developed to be used as attachment mechanism for further integration with robot arm. Nevertheless, further test will be conducted for this gripper such as the installment of force sensor, the measurement of grasping force and implementing force control during grasping operation.

6.0 References

[1] A. S. Sadun, J. Jalani, and J. A. Sukor, "An Overview of Active Compliance Control for a Robotic Hand," vol. 11, no. 20, pp. 2–5, 2013.

[2] A. G. Zisimatos, M. V Liarokapis, C. I. Mavrogiannis, and K. J. Kyriakopoulos, "Underactuated Robot Hands," no. Iros, pp. 3207–3212, 2014.

[3] B. Massa, S. Roccella, M. C. Carrozza, and P. Dario, "Design and development of an underactuated prosthetic hand," *Proc. 2002 IEEE Int. Conf. Robot. Autom. (Cat. No.02CH37292)*, vol. 4, no. May, pp. 3374–3379, 2002.

[4] T. Laliberté, L. Birglen, and M. C. Gosselin, "Underactuation in robotic grasping hands," *Mach. Intell. Robot. Control*, vol. 4, no. 3, pp. 1–11, 2003.

[5] H. Nishimura, A. Kakogawa, and S. Ma, "Development of an underactuated robot gripper capable of retracting motion," *2012 IEEE Int. Conf. Robot. Biomimetics, ROBIO 2012 - Conf. Dig.*, pp. 2161–2166, 2012.

[6] E. P. Puig, N. Eduardo, N. Rodriguez, and M. Ceccarelli, "A Methodology for the Design of Robotic Hands with Multiple Fingers," pp. 177–184, 2006.

[7] M. Plooij, G. Mathijssen, P. Cherelle, D. Lefeber, and B. Vanderborght, "Review of locking devices used in robotics," *IEEE Robot. Autom. Mag.*, vol. 22, no. 337596, pp. 106–117, 2010.

[8] A. S. Sadun, J. Jalani, and J. A. Sukor, "Model identification of a low-cost robot gripper by using matlab system identification toolbox (SIT) Model Identification Of A Low-Cost Robot Gripper By Using Matlab System Identification Toolbox (SIT)," no. January, 2016.

Acknowledgement

The authors would like to thank the Ministry of Education Malaysia for supporting this research under Fundamental Research Grant Scheme Vot No. FRGS/1/2018/TK05/UTHM/03/4 and partially sponsored by Universiti Tun Hussein Onn Malaysia.

IOP Conf. Series: Materials Science and Engineering **637** (2019) 012002 doi:10.1088/1757-899X/637/1/012002

Automatic fire fighting robot with notification

J Jalani[1], D Misman[1], A S Sadun[1] and L C Hong[1]

[1] Department of Electrical Engineering Technology (JTKE) Faculty of Engineering Technology (FTK) Universiti Tun Hussein Onn Malaysia
Email: jamalj@uthm.edu.my

Abstract. In real life, fire accidents can happen anytime and anywhere that usually hardly controllable. Fire accidents can damage the buildings, kill human and may cause unpredictable losses. In addition, death not only due to the fire but also from smoke inhalation and toxic gases. Therefore the fire security is important to human life. Note that the human is difficult to detect the fire in the location that is hard to reach or see by a human. Moreover, the human also may take a lot of time to extinguish fire due to the fact that finding a water source can be troublesome. Therefore, an automatic firefighting robot has been designed and proposed in this study. This robot used 3 flame sensors to detect the fire. It also equipped with 3 ultrasonic sensors for obstacle avoiding which protecting the robot and the internal components from any obstacles. Each sensor on the robot is controlled by the Arduino. Apart from the sensors, the robot is also equipped with the water tank that provides water once the fire is detected. The robot will move randomly in the room when the power is on. When the flame sensors detected the fire, the robot will move to the fire source and send a warning message to the user. Once the robot reached the burning area, it will stop at a certain distance and extinguish the fire by using water.

1. Introduction

The fire security of the home, office, and building is important to human life. A fast response to detect the small fire can avoid unpredictable damage and losses to human. However, it is difficult to detect the small fire in the location that is hard to reach or see by a human. The human also takes more time to find the water source to extinguish the fire. This will cause the fire to be spread quickly, which can increase death. Hence, to extinguish the fire within a short period of time and reduce the damage, an automatic fire fighting robot is proposed.

Robotics is one of the fastest-growing engineering fields. A lot of robots are designed to remove the human factor from labour-intensive or dangerous work. It also acts in an inaccessible environment. Today the use of robots is becoming more common than before and it is no longer exclusively used by the heavy production industries. A firefighting robot that is able to detect and extinguish the fire has been developed and used before. With the invention of such a device, people and property can be saved at a much higher rate with relatively minimal damage caused by the fire. To begin with, it is essential to design an autonomous robot and build a prototype system that could detect and extinguish fires automatically. The designed robot also able to move in a room with obstacle avoidance, detect the fire, extinguish it by using water and send a warning message to the user.

2. The design of the firefighting robot

Flame detectors represent one of the major types of automatic detection method and imitate the human sense of sight. They are the line of sight devices that operate on either an infrared, ultraviolet or combination principle. As radiant energy in the approximate 400 to 700 nanometers range occurs, as indicative of a flaming condition, the sensing equipment recognizes the fire signature and sends a signal to the fire alarm panel [1]. Kim et.al [2] had studied a fire alarm vision system based on flame detection algorithm of infrared images. From their studies, the presented flame detection algorithm measures the luminance variation of candidate flame blocks of the infrared image, then calculates the luminance variation value to determine the existence of fire. By using the test images, they performed the simulation of flame detection algorithms and tested the implemented fire alarm vision system.

The basic elements of this firefighting robot are the three flame sensors that monitor the temperature continuously. When the fire is detected, it is extinguished by water in the tank attached to the robot. But, if the temperature of the fire site is above 40°C, the alarm will be ringing so that the operator can control the firefighting robot to go back and avoid the damage of it [3]. This firefighting robot is designed to search for a fire in a small floor plan of a house, extinguish the fire and return to the front of the house at last [4]. In the target-driven obstacle avoidance model, the speed of a wheeled mobile robot is controlled by using fuzzy theory with sensor signals and allow it to move to a target location [5]. The ultrasonic sensor is the most suitable sensor for obstacle detection and avoidance because it is low cost and has a high range capability. The ultrasonic sensor is very compact and has a very high performance [6]. For the electronic circuit part, it consists Arduino Mega 2560, Adafruit motor shield V2, Ultrasonic Sensor, Flame sensor, HC-05 Bluetooth Module, 9V DC water pump, push-button, Lithium-ion Rechargeable Battery and AA Battery, 5V Single Channel Relay Module and a jumper wire. Table 1 shows the electrical components used in designing a circuit.

Table 1. Electric components for designing a circuit

Component	Features
Arduino Mega 2560	A microcontroller board based on the ATmega2560.
Adafruit motor shield V2	The motor shield is a driver module for motors that allows using Arduino to control the working speed and direction of the motor.
Ultrasonic Sensor	The ultrasonic ranging module HC-SR04 provides 2cm-400cm non-contact measurement function.
Flame sensor	The flame sensor is used to detect flame or infrared.
HC-05 Bluetooth Module	HC-05 Bluetooth module provides switching mode between master and slave mode which means it able to use neither receiving nor transmitting data.

The group in [7] had explored the implementation of Bluetooth technology in mobile robots. The mobile robot had the capability to move around autonomously by using a complicated and powerful algorithm. A PC stored the algorithms and act as a master cum server. The mobile robot will transmit all sensor readings to the master and processed. Subsequently, the server will transmit command or instruction for further action to the mobile robot in a bi-directional full-duplex communication mode. It focuses on the interfacing between Bluetooth transceiver and microcontroller of the mobile robot.

IOP Conf. Series: Materials Science and Engineering **637** (2019) 012002 doi:10.1088/1757-899X/637/1/012002

3. The architecture of the robot

There are two parts to be considered in the firefighting robot design. The first part is a mechanical design as shown in Figure 1. Meanwhile, the second part is the software design where the SketchUp software is used. The software is a 3D modeling, computer program for a wide range of drawing applications such as architectural, interior design, civil and mechanical engineering.

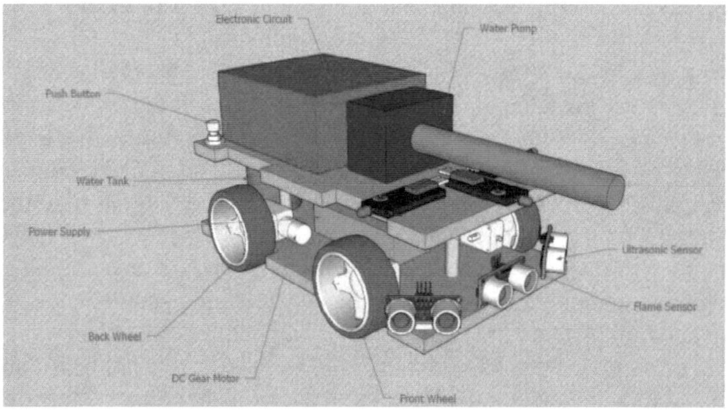

Figure 1. Design of the prototype.

3.1. Connection between Components

There are a lot of connections between components and Arduino by using jumper wires. The 4 DC gear motors were connected to the motor driver shield. All the components excluded DC gear motors are connected to an Arduino board as shown in Figure 2.

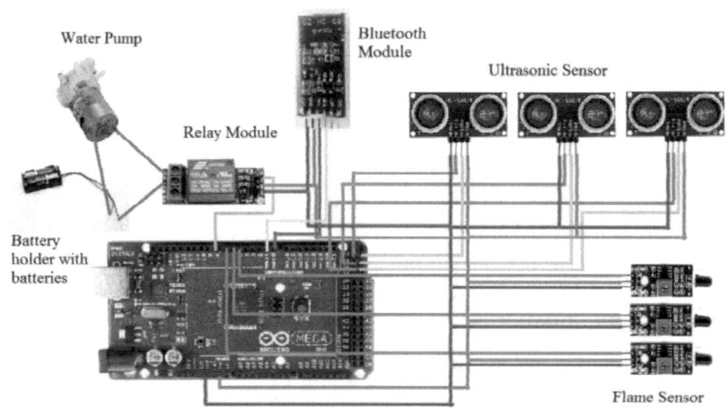

Figure 2. The connection between components and Arduino.

4. Arduino Software

In order to control the firefighting robot, Arduino Software was employed. Arduino programs may be written in any programming language with a compiler that produces binary machine code. The Arduino project provides the Arduino integrated development environment (IDE), which is a cross-platform application written in the programming language Java. The Arduino IDE supports the languages C and C++ using special rules to organize code. When the firefighting robot detecting the fire, it will send a warning notification to the user via phone by using Bluetooth module. Therefore, Arduino Wireless Vibration Apps was used to receive the signal from the Bluetooth module. If the flame sensor is triggered, it will send data "1" to android smartphone and giving a notification with sound and vibration on the Android smartphone by using this application. Vibrate interval and alarm sounds are adjustable in this application.

IOP Conf. Series: Materials Science and Engineering **637** (2019) 012002 doi:10.1088/1757-899X/637/1/012002

5. Results

Figure 3 shows the final configuration of an Automatic Fire Fighting Robot with notification. The control circuit is located at the top of the robot body. The size of the robot is (26 x 16 x 18) cm. The robot will operate in the auto mode after switch on, it will move randomly in house, office or any place that required early fire extinguisher. During the auto mode, if there is no fire detected, it will keep on moving and scanning. Meanwhile, if fire detected, it will move toward the fire source and activate the water pump to pump the water to extinguish the fire. If the left flame sensor detecting fire, the left sensor will send the data to Arduino telling that the fire is at the left-hand side, the Arduino will send the command to the motor driver to turn the robot to left direction. The robot will remain turning to left direction until the front flame sensor detecting fire. When the front sensor detecting fire, the sensor will send the data to Arduino to tell the Arduino that there is a fire in front of it. Then the Arduino will send data to the motor driver shield and motor driver will activate the motors to move the robot backward the fire if the fire is too close to the robot. Otherwise, if the fire is too far from the robot, the motor driver will activate the motors to move the robot forward to the fire. This process will be the same if the right flame sensor detecting fire.

Figure 3. Automatic Fire Fighting Robot with Notification

6. Analysis

6.1. Fire Detection and Extinguishing

The flame sensors will only activate to detect fire when the robot is powered on. The operating voltage for the flame sensor is 5V. Voltage supply for theses sensors is from Arduino that supplied by using 7.4V Li-ion rechargeable battery. The voltage will decrease when the sensors detect fire. The voltage will decrease from 4.8V to 0.1V, if the voltage continues to decrease until reached 0.1V, the sensor will recognize the input as "fire" and send digital data to Arduino. The Arduino will determine which sensors are sending the data and send commands to the motor driver for activated the DC motor, Hence the DC motor triggered according to which sensor detecting fire first. The robot will move toward the fire until it reached extinguished range set in the program. When the robot reached the extinguished range, it will start to pump the water from the water tank.

6.2. Sending Notification by Bluetooth Module

"Automatic Fire Fighting Robot with Notification" able to send a warning notification to the user when the robot detecting the fire. This can alert the user which action can be immediately taken to stop the fire by calling the firefighter. Of course, when action is taken by the user, the firefighting robot has performed its function.

Figure 4. Bluetooth module and inactive alarm

Figure 5. Bluetooth module and active alarm

This operation was carried out by using a Bluetooth module. Bluetooth module inside the robot able to send the notification to the user via smartphone after the robot detecting fire. When the user's smartphone received the signal from the Bluetooth module, it will vibrate and sound. Therefore, the user knows that there is a fire accident and able to immediately take action to control the fire. Figure 4 and figure 5 show the Bluetooth module and the alarm when the flame sensor detecting fire.

7. Conclusion

The automatic fire fighting robot with notification was able to detect and extinguish the fire. Besides, it is also able to move randomly in a room with obstacle avoidance function. The robot was successfully fabricated and functioned as expected. The hardware was developed according to the initial design and modified based on current conditions and improvements. In addition, the input voltage port and ground port of the sensors, Bluetooth module and relay was connected and soldering on the PCB in series. This is because there is only one port of 5V voltage supply from Arduino, but there are 8 components were required for the voltage supply for operation. Next, the coding was designed and developed using Arduino software. The completeness of the software library smoothens the project's process with the help of real-time simulation. However, it was observed that the accuracy of the fire extinguisher can be further improved despite the firefighting robot designed and presented here provides the significance assistance to human. This firefighting robot concept can be widely applied in various applications.

8. References

[1] An Introduction to Fire Detection, Alarm, and Automatic Fire Sprinklers -NEDCC. [Online]. Available: https://www.nedcc.org/freeresources/preservation-leaflets/3.-emergency-management/3.2-anintroduction-to-fire-detection,-alarm,-and-automatic-fire-sprinklers. [Accessed: 22-May-2016].
[2] Kim W, Kim S, Lee J, and Hyun C A Fire Alarm Vision System based on IR Image Processing no. 1 pp 291–293
[3] Aung P W and Win W Y 2014 Remote Controlled Fire Fighting Robot vol 03 no 24 pp 4830–4835.
[4] Parana U J S and Prasad M V D 2013 Automatic Fire Sensing and Extinguishing Robot Embedded With GSM Modem no 4 pp 221–224.
[5] Chen Y and Juang J 2009 Intelligent Obstacle Avoidance Control Strategy for Wheeled Mobile Robot ICROS-SICE Int. Jt. Conf. pp 3199–3204.
[6] Sonali K K, Dharmesh H S and Nishant M R 2010 Obstacle avoidance for a mobile exploration robot using a single ultrasonic range sensor Interact. pp 8–11.
[7] Choo S H, Amin S H M, Fisal N, Yeong C F and Abu Bakar J 2002 Using Bluetooth transceivers in mobile robot Student Conf. Res. Dev. pp 472– 476.

Acknowledgment

The research leading to these results has received funding from the Registrar of the University Tun Hussein Onn Malaysia (UTHM). The authors also wish to thank the Faculty of Engineering Technology, UTHM for providing a platform to carry out this research.

IOP Conf. Series: Materials Science and Engineering **637** (2019) 012003 doi:10.1088/1757-899X/637/1/012003

A Novel Robust Adaptive and Compound Control of an Adaptive Neural Network, SMC and PI for Manipulators

Duc Ha Vu[1,2], Shoudao Huang[1] and Thi Diep Tran[1,2]

[1] College of Electrical and Information Engineering, Hunan University, Hunan, China
[2] Faculty of Electrical Engineering, Saodo University, Haiduong, Vietnam
Corresponding author's e-mail address: vuhadhsd@hnu.edu.cn

Abstract. In this paper, a new compound control scheme is proposed for manipulator based on radial basis function neural network (RBFNNs), sliding mode controller (SMC) and proportional–integral (PI) controller. In this control scheme, the filtered tracking error is the input of the RBFNNs update laws, SMC, and PI controller. The RBFNNs uses three-layer to approximate uncertain nonlinear manipulator dynamics. A robust sliding function is selected as a second controller to guarantee the stability and robustness under various environments. By using additional PI controllers, the goal of manipulator tuning is to minimize chattering signal, tracking performance, and overshoot can be realized. Simulation results highlight performance of the controller to compensate the approximate errors and its simpleness in the adaptive parameter tuning process. To be concluded, the controller is suitable for robust adaptive control and can be used as supplementary of traditional neural network (NN) controllers.

1. Introduction

Artificial neural networks, as a kind of tools of computational intelligence, have been credited in various applications as powerful tools capable of providing robust controllers for mathematically ill-defined systems that may be subjected to structured and unstructured uncertainties [1]. The universal approximation theorem has been the main driving force behind the increasing popularity of such methods, as it shows that they are theoretically capable of uniformly approximating a continuous real function to any degree of accuracy. This has led to recent advances in the area of intelligent control [2]. The properties of NN are learning, nonlinear mapping, generalization abilities and parallelism of computation. Because of these characteristics, NN becomes an important application in areas such as identification, optimization, intelligence control, robotics, etc. NN can provide online learning algorithms and deal with unmodeled unknown dynamics in manipulator model. These algorithms are designed on the basis of the Lyapunov stability theorem. In [3], the authors proposed a control compensator using a neural network to compensate for the control errors of the control system. This controller has proposed a new command control based on tracking error of the desired position trajectories and actual position system when the parameter changes and large disturbances. In [4], the authors researched the tracking problem based on the SBLF and adaptive Neural Networks. The proposed controller was combined backstepping and adaptive feedback approximation method to improve the control performance. In [5], Bin Xu, Chengguang, and Zhongke Shi were proposed an adaptive critic based NN controller to solve the nonlinear systems. Both the tracking error and strategic utility function were improved by using the action neural network to approximate the nonlinear functions. In [6], an adaptive NN controller was suggested for an uncertain robot with unknown dynamics. The authors employed the adaptive neural networks to approximate the unknown model of the robot, shut the door on the violation of the constraints, and compensate for the unknown of the dynamic structure of the robot. In [7], the authors have been studied the tracking problem of 3-DOF robotic based on an adaptive neural network controller. This controller has been considered both output feedback and state control schemes by using two neural networks. A neural network was employed to approximate the dynamic of the robot and another neural network uses to approximate the unknown hysteresis non-linearity. In [8], a robust adapted controller on the basis of neural networks was introduced to control the SCARA robot arm. In [9] an adaptive output feedback control method uses the proposed RBF network to adaptive compensate for the tracking output of continuous nonlinear systems. In [10] an adaptive tracking control scheme based on the neural network is proposed for a class of nonlinear systems. In which RBF neural network is used to adaptive learn the uncertainty bounded of the system according to the Lyapunov stability theorem, and the output of the neural network is used as the parameters of the controller to compensate uncertainty impacts of the system. In the above documents have inherited advantages of the neuron controller, which is the ability to approximate and learning the rules online during the controller work. However, the neuron controller cannot avoid approximate errors. To solve the disadvantages of the neuron controller, it is necessary to provide a combination controller between the neuron controller and the sliding controller [11, 12]. In which, the authors have combined the neuron controller with the sliding controller to control industrial robot manipulators. The neuron controller with fast learning algorithm and good approximation ability and the sliding controller with robust compensating effect acts as a secondary controller to ensure stability and sustainability under the different environments. These two studies both achieve the correct tracking performance and high stability of the controller. However, the control signal in [11, 12] still has chattering phenomenon. To overcome the above disadvantages, in the author's paper, not only does the combination SMC with the neuron network (NN) to compensate the approximation error, but it also incorporates a PI controller to improve chattering signal, tracking performance and overshoot. Therefore different with existing control methods, this method combines the RBFNNs, SMC and the PI to improve the robust ability, control performance and thus can be seemed like a robust adaptive controller. As shown in the simulation results, when applying this controller to control a two-link robot manipulator compared with the RBFNNs [12], PD the performance of the proposed control is improved so much.

2. Dynamics of manipulators

Consider the dynamic equation of the n-link rigid manipulator can be written as:

$$M(q)\ddot{q} + C(q,\dot{q})\dot{q} + G(q) + F(\dot{q}) = \tau - T_d \tag{1}$$

Where $(q,\dot{q},\ddot{q}) \in R^{n\times 1}$ are the vectors of joint position, angle velocity and angle acceleration, respectively. $M(q) \in R^{n\times n}$ is the symmetric inertial matrix. $C(q,\dot{q}) \in R^{n\times n}$ is the vector of Coriolis and Centripetal forces. $G(q) \in R^{n\times 1}$ expresses the gravity vector. $F(\dot{q}) \in R^{n\times 1}$ represents the vector of the frictions $\tau \in R^{n\times 1}$ is vector of input torques. $T_d \in R^{n\times 1}$ is vector of disturbance torques. For the purpose of designing controller, several properties of the manipulator model (1) have been assumed as follows:

Property 1: The inertial matrix $M(q)$ is a positive symmetric matrix and is defined by:

$$m_1\|x\|^2 \le x^T M(q)\mathrm{x} \le m_2\|x\|^2,\ \forall x \in R^n \tag{2}$$

with m_1 and m_2 are known positive constants and they depend on the mass of the manipulators.

Property 2: $\dot{M}(q) - 2C(q,\dot{q})$ is skew symmetry matrix, in which

$$x^T\left[\dot{M}(q) - 2C(q,\dot{q})\right]x = 0 \tag{3}$$

Property 3: $C(q,\dot{q})\dot{q}$, $F(\dot{q})$ and $G(q)$ are bounded as follows:

$$\|C(q,\dot{q})\dot{q}\| \le C_k\|\dot{q}\|^2,\ \|F(\dot{q})\| \le F_k\|\dot{q}\| + F_0, \|G(q)\| \le G_k \tag{4}$$

with C_k, G_k, F_k, F_0 are positive constants.

3. Control strategy

Control strategy based on design of RBFNNs controller, SMC and the PI controler. The RBFNNs controller is designed based on the structure of radial basis function (RBF) neural networks. After analyzing the structure of RBF neural networks and the dynamics of manipulators, a new compound control scheme is proposed.

3.1. The structure of RBF neural networks

RBF neural network is a type of network capable of approximation any continuous nonlinear functions with arbitrary accuracy by the number of neurons in the hidden layer. Furthermore, RBFNNs controllers have been successfully deal with manipulators control problems mentioned by many researchers. In this report, the RBF neural network proposed in [11] is selected as the network controller for its Lyapunov stability. Assume that $M(q),C(q,\dot{q}),G(q)$ and $F(\dot{q})$ the output values of ideal RBFNNs and determined, respectively as

$$M(q)=M_R(q)+\varepsilon_M=W_M^T*h_M(q)+\varepsilon_M \quad (5)$$

$$C(q,\dot{q})=C_R(q,\dot{q})+\varepsilon_C=W_C^T*h_c(q,\dot{q})+\varepsilon_C \quad (6)$$

$$G(q)=G_R(q)+\varepsilon_G=W_G^T*h_G(q)+\varepsilon_G \quad (7)$$

$$F(\dot{q})=F_R(\dot{q})+\varepsilon_F=W_F^T*h_F(\dot{q})+\varepsilon_F \quad (8)$$

With W_M,W_C,W_G and W_F are ideal optimum weight value of RBF; h_M,h_C,h_G and h_F are outputs of hidden layer, $\varepsilon_M,\varepsilon_C,\varepsilon_G$ and ε_F are modeling error of $M(q),C(q,\dot{q}),G(q)$ and $F(\dot{q})$, respectively. $\hat{M}_R(q),\hat{C}_R(q,\dot{q}),\hat{G}_R(q)$ and $\hat{F}_R(\dot{q})$ are the estimated values of the $M_R(q),C_R(q,\dot{q}),G_R(q)$ and $F_R(\dot{q})$, respectively. They are described by RBF as follows:

$$\hat{M}_R(q)=\hat{W}_M^T*h_M(q) \quad (9)$$

$$\hat{C}_R(q,\dot{q})=\hat{W}_C^T*h_C(q,\dot{q}) \quad (10)$$

$$\hat{G}_R(q)=\hat{W}_G^T*h_G(q) \quad (11)$$

$$\hat{F}_R(\dot{q})=\hat{W}_F^T*h_F(\dot{q}) \quad (12)$$

3.2. Compound control scheme

Robotic manipulator dynamics with multi-joints are highly nonlinear, highly coupling and form an uncertain model system. The RBFNNs controller has many advantages, which is the ability to approximate and learning the rules online during the controller work. However, the neuron controller cannot avoid approximate errors. As a result, a compound control method based on RBFNNs, SMC and PI is introduced for the position-tracking control of manipulators. The architecture of the manipulator control system is shown in Fig. 1.

Determine a tracking error vector $e(t)$ and the filtered tracking error $s(t)$ as the following equations:

$$e(t)=q_d-q \quad (13)$$

$$s(t)=\dot{e}-\lambda e \quad (14)$$

where $\lambda=\lambda^T>0$, differentiating $s(t)$ and using (1), the manipulators dynamics may be written in terms of the sliding mode functions as:

$$M\dot{s}=-Cs+M(\ddot{q}_d+\lambda\dot{e})+C(\dot{q}_d+\lambda e)+G+F+T_d-\tau \quad (15)$$

Since equation (5) - (8), (15) becomes: $M\dot{s}=-Cs+f(x)-\tau+T_d+\varepsilon$ (16)

Where $f(x)$ is defined as $f(x)=M_R(\ddot{q}_d+\lambda\dot{e})+C_R(\dot{q}_d+\lambda e)+G_R+F_R$ and

$$\varepsilon=\varepsilon_M(\ddot{q}_d+\lambda\dot{e})+\varepsilon_C(\dot{q}_d+\lambda e)+\varepsilon_G+\varepsilon_F+T_d$$

For the dynamics of an n-link robot manipulator (1), the compound control law is proposed as:

$$\tau = \hat{f}(x) + \tau_s + \tau_{PI} \tag{17}$$

where: $$\tau_{PI} = K_P s + K_I \int_0^{\tau} s dt,\ K_P = diag\{K_{p1}, K_{p2}, ..., K_{pn}\},\ K_I = diag\{K_{i1}, K_{i2}, ..., K_{in}\} \tag{18}$$

K_P, K_I is the positive definite matrix, τ_s is a SMC robust term that is used to suppress the effects of uncertainties and approximation errors, and $\hat{f}_x$ is the approximation of the adaptive function f_x and is defined as: $\hat{f}(x) = \hat{M}_r(\ddot{q}_d + \lambda\dot{e}) + \hat{C}_R(\dot{q}_d + \lambda e) + \hat{G}_R + \hat{F}_R$

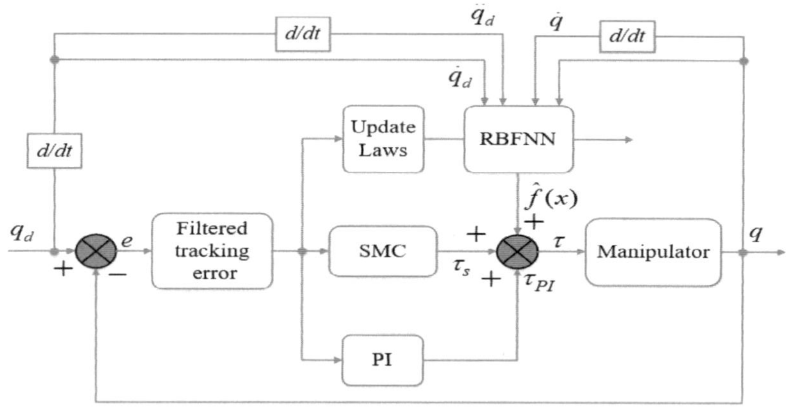

Figure 1. Proposed control scheme.

Follow above analysis, we propose a SMC robust term τ_s as:

$$\tau_s = K_s \operatorname{sgn}(s) \tag{19}$$

where $K_s = diag\{K_{s1}, K_{s2}, ...K_{sn}\}$, and $K_s > \|\varepsilon\|$.

Substituting (17) into (16), we have:

$$M(q)\dot{s} = -C(q,\dot{q})s + \tilde{f}(x) - \tau_{PI} - \tau_s + T_d + \varepsilon \tag{20}$$

where

$$\tilde{f}(x) = f(x) - \hat{f}(x) = \tilde{M}_R(\ddot{q}_d + \lambda\dot{e}) + \tilde{C}_R(\dot{q}_d + \lambda e) + \tilde{G}_R(q) + \hat{F}_R \tag{21}$$

By applying the adaptive control law equation (17) to the dynamic equation (1), using the SMC function equation (19), and using the PI controller equation (18), the online RBF neural networks adaptive update laws are designed as:

$$\begin{cases} \dot{\hat{W}}_M = k_M \Xi_M s - \Xi_M h_M s^T \\ \dot{\hat{W}}_C = k_C \Xi_C s - \Xi_C h_C s^T \\ \dot{\hat{W}}_G = k_G \Xi_G s - \Xi_G h_G s^T \\ \dot{\hat{W}}_F = k_F \Xi_F s - \Xi_F h_F s^T \end{cases} \tag{22}$$

where Ξ_M, Ξ_C, Ξ_G và Ξ_F are symmetric positive definite constant matrices. k_M, k_C, k_G and k_F are positive adaptation rates.

The stability of the closed-loop system in Fig. 1 which is described by equation (17) is established in the following theorem.

Theorem: Consider an n-link manipulator represented by (1). If the RBFNNs adaptive update laws are designed as (22), the SMC is give by (19), and PI controler (18). The adaptive control law designed in (17), then the tracking error and the convergence of all the system parameters can approached to zero asymptotically.

Proof: Consider the following candidate Lyapunov function:

$$V(t)=\frac{1}{2}\begin{bmatrix} s^T M s+(\int_0^t s dt)^T K_I (\int_0^t s dt)+tr(\tilde{W}_M^T \Xi_M^{-1} \tilde{W}_M) \\ +tr(\tilde{W}_C^T \Xi_C^{-1} \tilde{W}_C)+tr(\tilde{W}_G^T \Xi_G^{-1} \tilde{W}_G)+tr(\tilde{W}_F^T \Xi_F^{-1} \tilde{W}_F) \end{bmatrix} \tag{23}$$

where $\tilde{W}_M = W_M - \hat{W}_M ; \tilde{W}_C = W_C - \hat{W}_C ; \tilde{W}_G = W_G - \hat{W}_G ; \tilde{W}_F = W_F - \hat{W}_F$.

The derivative of $V(t)$ along to time, we have:

$$\dot{V}(t)=s^T\left(M\dot{s}+\frac{1}{2}\dot{M}s+K_I\int_0^t s dt\right)+tr(\tilde{W}_M^T \Xi_M^{-1} \dot{\tilde{W}}_M)+tr(\tilde{W}_C^T \Xi_C^{-1} \dot{\tilde{W}}_C)+tr(\tilde{W}_G^T \Xi_G^{-1} \dot{\tilde{W}}_G)+tr(\tilde{W}_F^T \Xi_F^{-1} \dot{\tilde{W}}_F) \tag{24}$$

Submitting (20) into (24), yields

$$\begin{aligned} \dot{V}(t) &= -s^T K_P s+\frac{1}{2}s^T(\dot{M}-2C)s+s^T\varepsilon - s^T\tau_s + tr(\tilde{W}_M^T \Xi_M^{-1} \tilde{W}_M + h_M s^T) \\ &+ tr(\tilde{W}_C^T \Xi_C^{-1} \tilde{W}_C + h_C s^T) + tr(\tilde{W}_G^T \Xi_G^{-1} \tilde{W}_G + h_G s^T) + tr(\tilde{W}_F^T \Xi_F^{-1} \tilde{W}_F + h_F s^T) \end{aligned}$$

By using property 2, and since $\dot{\tilde{W}} = -\dot{\hat{W}}, s^T \tilde{W}^T h(x) = tr(\tilde{W}^T h(x) s^T)$, and from adaptive law (22), (24) becomes:

$$\begin{aligned} \dot{V}(t) &= -s^T K_P s + s^T\varepsilon - s^T\tau_s + str\left[\tilde{W}_M^T(k_M\hat{W}_M)\right] + str\left[\tilde{W}_C^T(k_C\hat{W}_C)\right] + str\left[\tilde{W}_G^T(k_G\hat{W}_G)\right] + str\left[\tilde{W}_F^T(k_F\hat{W}_F)\right] \\ &= -s^T K_P s + s^T\varepsilon - s^T K_s sign(s) + s\left\{tr\left[k_M\tilde{W}_M^T(W_M-\tilde{W}_M)\right] + tr\left[k_C\tilde{W}_C^T(W_C-\tilde{W}_C)\right]\right. \\ &\left. + tr\left[k_G\tilde{W}_G^T(W_G-\tilde{W}_G)\right] + tr\left[k_F\tilde{W}_F^T(W_F-\tilde{W}_F)\right]\right\} \end{aligned} \tag{25}$$

By using $tr\tilde{W}(W-\tilde{W}) = (\tilde{W},W) - \|\tilde{W}\|^2 \le \|\tilde{W}\|\|W\| - \|\tilde{W}\|^2$ and $\|W\| \le W_{\max}$

We have

$$\begin{aligned} \dot{V}(t) &\le -s^T K_P s + s^T\varepsilon - s^T K_s sign(s) + k_M s\left(\|\tilde{W}_M\|\|W_M\| - \|\tilde{W}_M\|^2\right) + k_C s\left(\|\tilde{W}_C\|\|W_C\| - \|\tilde{W}_C\|^2\right) \\ &+ k_G s\left(\|\tilde{W}_G\|\|W_G\| - \|\tilde{W}_G\|^2\right) + k_F s\left(\|\tilde{W}_F\|\|W_F\| - \|\tilde{W}_F\|^2\right) \\ &\le -s^T K_P s + s^T\varepsilon - s^T K_s sign(s) + k_M s\left[\left(\|\tilde{W}_M\| - \frac{W_{M\max}}{2}\right)^2 - \frac{W_{M\max}^2}{4}\right] \\ &+ k_C s\left[\left(\|\tilde{W}_C\| - \frac{W_{C\max}}{2}\right)^2 - \frac{W_{C\max}^2}{4}\right] + k_G s\left[\left(\|\tilde{W}_G\| - \frac{W_{G\max}}{2}\right)^2 - \frac{W_{G\max}^2}{4}\right] \\ &+ k_F s\left[\left(\|\tilde{W}_F\| - \frac{W_{F\max}}{2}\right)^2 - \frac{W_{F\max}^2}{4}\right] \end{aligned} \tag{26}$$

In (26), considering $K_s > \|\varepsilon\|$, and if gain K_P and s are selected to satisfy the following inequality:

$$K_P \ge \frac{1}{s}\left[\frac{k_M W_{M\max}^2}{4} + \frac{k_C W_{C\max}^2}{4} + \frac{k_G W_{G\max}^2}{4} + \frac{k_F W_{F\max}^2}{4}\right] \tag{27}$$

Then

$$\dot{V}(t) \le 0 \tag{28}$$

Since (28), $\dot{V}(s(t),\tilde{W}_M,\tilde{W}_C,\tilde{W}_G,\tilde{W}_F) \le 0$ is a negative semidefinite function, $\dot{V}(s(0),\tilde{W}_M,\tilde{W}_C,\tilde{W}_G,\tilde{W}_F)$ $\le \dot{V}(s(0),\tilde{W}_M,\tilde{W}_C,\tilde{W}_G,\tilde{W}_F)$ if all parameters such as $s(t),\tilde{W}_M,\tilde{W}_C,\tilde{W}_G,\tilde{W}_F$ arebounded with $t>0$. By defining $\Gamma(t) = s^T K_p s - s^T\varepsilon + s^T K_s sign(s)$, so $\Gamma(t) \le -\dot{V}(t)$ and integrate the $\Gamma(t)$ with respect to time as follows: $\int_0^t \Gamma(\zeta)d\zeta \le V(s(0),\tilde{W}_M,\tilde{W}_C,\tilde{W}_G,\tilde{W}_F) - V(s(t),\tilde{W}_M,\tilde{W}_C,\tilde{W}_G,\tilde{W}_F)$.

Because $V(s(0),\tilde{W}_M,\tilde{W}_C,\tilde{W}_G,\tilde{W}_F)$ is a bounded function, and $V(s(0),\tilde{W}_M,\tilde{W}_C,\tilde{W}_G,\tilde{W}_F)$ is nonincreasing and bounded, we have:

$$\lim_{t\to\infty}\int_0^t \Gamma(\zeta)d\zeta < \infty \tag{29}$$

According to Barbalat's Lemma [13], when $\Gamma(t)$ is bounded function. It can be shown that $\lim_{t\to\infty}\int_0^t \Gamma(t)dt = 0$. From this result, we see that $s(t)$ will converge to zero when $t \to \infty$ and the global stability of the control system for the manipulator is guaranteed by the compound control law (17).

4. Simulation results and discussions

In order to show the effectiveness of the proposed control system, simulation experiments are carried out. In this section, we consider a two-link robot manipulator. The dynamics of the manipulator can be obtained as in [12]. In the simulation experiments, the desired position trajectories are chosen by: $q_{1d} = q_{2d} = 0.5\sin(t)$, and initial positions of joints are $q_0 = \begin{bmatrix}0.09 & -0.09\end{bmatrix}^T$, and initial velocities of joints are $\dot{q}_0 = \begin{bmatrix}0.0 & 0.0\end{bmatrix}^T$. The parameter values used in the compound control system are:
$\Xi_M = \Xi_C = \Xi_G = \Xi_F = diag[15, 15]; \lambda = diag[5,5]; K_P = diag[50,50]; K_I = diag[100,100];$
$K_s = diag[0.1,0.1]$.

In the following passage, our proposed neuron network control scheme is applied to comparison with the RBFNNs [12] and the proportional differential (PD) controller, where the output of the PD controller can be expressed to be $y_{pd} = K\chi e(t) + K_\delta \dot{e}(t)$, and the PD gains were selected to be $K\chi = diag[200, 250]; K_\delta = diag[20, 20]$.

Fig. 2 shows the position tracking trajectories of the two joints by using the RBFNNs, PD and the proposed controller. Fig. 3 presents the tracking errors of the two joints by using the RBFNNs, PD and the proposed controller. In Fig. 4, the control efforts of the RBFNNs, PD and the proposed controller is presented. To validate ability minimize chattering signal, control input of the RBFNNs, PD and the proposed controller for zoomed-in time frame (5 – 7 s) can see clearly in Fig 5.

The above simulation results for the two-link manipulator demonstrate that the PD control has low trajectory tracking precision, slow response time and large position-tracking errors. This is because the PD controller only controls for objects with nominal models. The PD controllers are unable to determine the appropriate PD gains in the case of nonlinear and uncertain controlled plants and can't compensate static errors.

As shown in Fig. 2, 3 and 4 the RBFNNs and the proposed controller can realize higher trajectory tracking precision and better adaptive control capability due to the powerful capabilities of learning adaptability. This is also because in the RBFNNs controller and the proposed controller contains a robust component. It is the SMC that compensates approximate errors in the neuron network controller. However, see in Fig. 4, 5 the SMC causes in the RBFNNs is chattering phenomenon for the control force. To solve the above problem, in our controller with simple and more effective parameter update law. Our update laws only have four parameters to update while the update laws in RBFNNs have six parameters to update. This leads to our controller being able to calculate and simulate faster. In addition, by using additional PI controllers, the goal of the proposed controller is to minimize chattering signal, error tracking, and overshoot is achieved when compared with the RBFNNs, can see in Fig. 2, 3, 4 and 5. This is because proportional gain (K_P) works in conjunction with integral gain (K_I) to reduce overshoot, tracking error and provide damping to the system, while keeping response time to acceptable levels. The advantages of the proposed control scheme have position tracking improvements better than that of the RBFNNs, PD and all updated parameters in the dynamic structure

RBFNNs are simpleness adjusted. Moreover, from Fig. 4, 5 can observe that the control force of the proposed controller is smoother and has smaller oscillation than the RBFNNs and PD to achieve the requested level of performance when the tracking errors reach the big value.

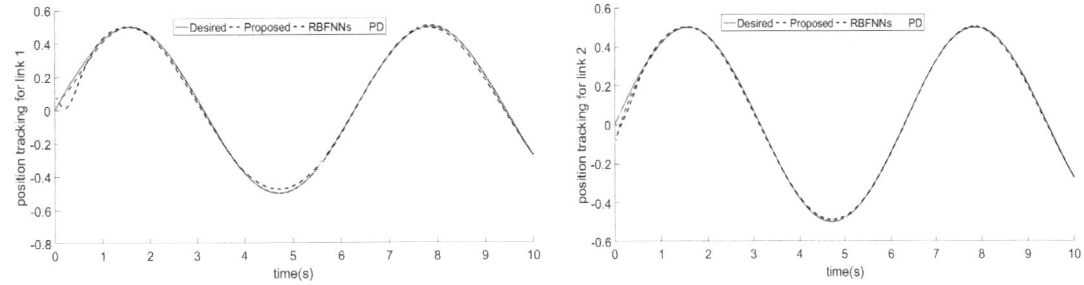

(a) (b)

Figure 2. Position responses of the RBFNNs, PD and the proposed controller: (a) link-1; (b) link-2.

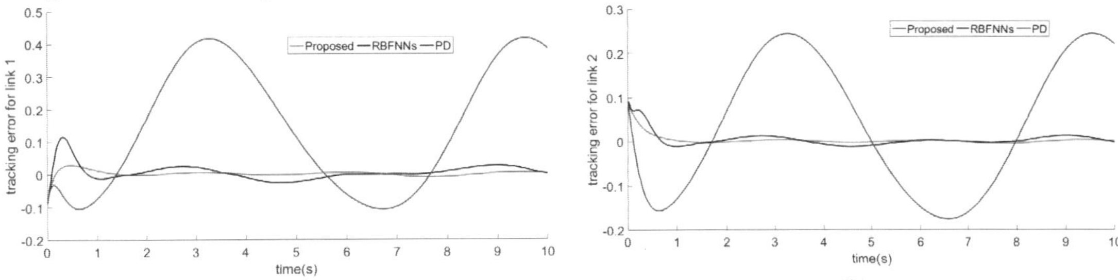

(a) (b)

Figure 3. Tracking errors of the RBFNNs, PD and the proposed controller: (a) link-1; (b) link-2.

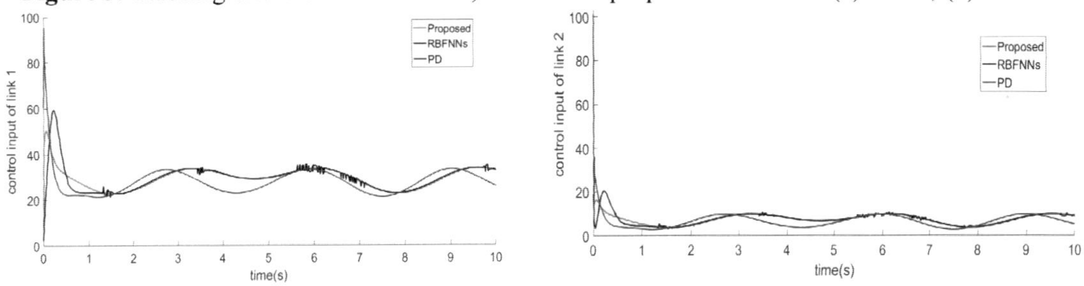

(a) (b)

Figure 4. Control efforts of the RBFNNs, PD and the proposed controller: (a) link-1; (b) link-2.

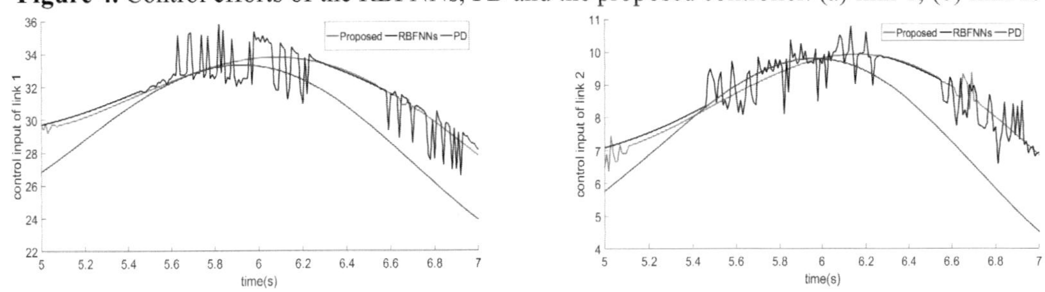

(a) (b)

Figure 5. Control efforts of the RBFNNs, PD and the proposed controller for zoomed-in time frame (5–7 s): (a) link-1; (b) link-2.

5. Conclusions

In this paper, a new compound NN, SMC and PI control scheme has been proposed. It has been also successfully implemented to control the joints of a two-link robot manipulator for achieving high precision position tracking by combining the advantages of radial basis function neural network, sliding mode robust term function and PI controller to improve chattering signal, tracking performance and overshoot. The difficulty to find approximate values of the unknown dynamic of manipulator has been to solve by radial basis function neural network control. All the adaptive online training for the weights of the radial basis function neural network are obtained by Lyapunov theorem, and trained online by an adaptive learning algorithm. From the simulation results of a two-links robot manipulator, we can see that the performance of the proposed control is improved so much. The future research work, we can continue to research to put into experimental as well as be applied in practice.

6. Reference

[1] Chaoui H and Gueaieb W 2008 Type-2 fuzzy logic control of a flexible-joint manipulator *Journal of Intelligent and Robotic Systems* **51** 159-86
[2] Sun C, He W and Hong J 2017 Neural network control of a flexible robotic manipulator using the lumped spring-mass model *IEEE Transactions on Systems, Man, and Cybernetics: Systems* **47** 1863-74
[3] Choi Y K, Lee M J, Kim S and Kay Y C 2001 Design and implementation of an adaptive neural-network compensator for control system *IEEE Transactions on Indus-trial Electronics* **48** 416-23
[4] Zhao Z, He W and Ge S S 2014 Adaptive neural network control for a fully actuated marine surface vessel with multiple output constraints *IEEE Trans. Control Systems Technology* **22** 1536-43
[5] Xu B, Yang C G and Shi Z K 2014 Reinforcement learn-ing output feedback NN control using deterministic learn-ing techniques *IEEE Trans. Neural Network and Learn-ing Systems* **25** 635-41
[6] He W, Chen Y and Zhao Y 2016 Adaptive neural network control of an uncertain robot with full-state constraints *IEEE Transactions on Cybernetics* **46** 620-29
[7] He W, David O A, Yang C and Dawei G 2017 Adaptive neural network control of a robotic manipulator with un-known backlash-like hysteresis *IET Control Theory Appl* **11** 567-75
[8] Rossomando F G and Soria C M 2016 Adaptive neural sliding mode control in discrete time for a SCARA robot arm *IEEE Latin america transactions* **14** 2556-64
[9] Seshagiri S and Khail H K 2000 Output feedback control of nonlinear systems using RBF neural networks *IEEE Trans. Neural Networks* **11** 69–79
[10] Zhihong M, Wu H R and Palaniswami M 1998 An adaptive tracking controller using neural networks for a class of nonlinear systems *IEEE Trans. Neural Networks* **9** 947–55
[11] Yen V T, Nan W Y and Cuong P V 2019 Robust Adaptive Sliding Mode Neural Networks Control for Industrial Robot Manipulators *International Journal of Control, Automation and Systems* **17** 783-92
[12] Cuong P V and Nan W Y 2015 Adaptive trajectory tracking neural network control with robust compensator for robot manipulators *Neural Computing and Applications* **27** 525-36
[13] Slotine J J E and Li W 1991 *Applied Nonlinear Control* (Hoboken: Prentice-Hall)

IOP Conf. Series: Materials Science and Engineering **637** (2019) 012004 doi:10.1088/1757-899X/637/1/012004

Discussion on the welding performance of automatic welding of industrial robot and manual welding

ZHANG Yan

Huanghebei Road,Xiyanduan district, Xuzhou City, Jiangsu Province, China
Email: zyredeemer@outlook.com

Abstract. Since the introduction of industrial robots and their widespread application in manufacturing, the production quality and efficiency of manufacturing have been affected their automated welding performance directly. In this paper the author provides a useful basis for improving welding quality and promoting production efficiency through the analysis of industrial robot automatic welding technology standards and welding performance.

Keywords: automated welding, plat butt welding, tensile test for deposited metal, microstructure analysis

1. Introduction

Welding is a widely used processing technology in the manufacturing industry, with the advantages of low cost, good performance of connection and easy operation, etc. The introduction of industrial robots undoubtedly injected fresh blood into the manufacturing industry although the global manufacturing industry continues to fall now, and the quality of manufacturing products is affected by the level of welding technology directly. As technology continues to evolve, automated welding techniques are becoming more sophisticated, but there are still questions about whether the welding performance of automated welding is sufficient to meet needs of production. In the process of cooperation and exchange with XuZhou Construction Machinery Group, the author compares the tensile strength test of welded plates, the tensile and impact tests of deposited metal and the structural analysis of T-joint welds to compare the performance strength of robot welding and manual welding.

2. Technical standard of industrial robotic automatic welding

2.1. Principles of industrial robot automatic welding

At present, most of the intelligent positioning techniques of welding robots use wire contact positioning, and some welding robots use nozzle contact positioning [1].

IOP Conf. Series: Materials Science and Engineering **637** (2019) 012004 doi:10.1088/1757-899X/637/1/012004

The robot contact positioning circuit is a normally open circuit. When the welding wire or nozzle contacts the workpiece, the open circuit is closed and the built-in relay of the robot is closed, then the processor records the position information of the contact point. During the positioning process, the deviation between the actual position of the workpiece surface reference point and the programmed position can be determined and the corresponding welding procedure is corrected. When the welding contact positioning has an error, the position of the weld automatically corrected by the robot will also change [2].

2.2. Methods and standards for automated welding

2.2.1. Two smooth polished samples which have a 60° slope is reserved for the butt assembly of the workpiece are placed and fixed on the tooling table with a clamp. Make sure there are no people and foreign objects in the robot's working range. Run the clearing program to check whether the robot's TCP point deviates and the dry elongation is between 18 and 22 mm after the reduction of the wire. If the TCP point deviates, check if the contact nozzle is worn and the torch, clamp holder, and sway bar are deformed. After correcting the TCP point, the cleaning procedure should be debugged to ensure that the gun is safe and reliable during the cleaning process, and the dry extension is between 18 and 22 mm after the gun is cleaned and the wear is reduced.

2.2.2. Turn on the wire signal, pull the wire with a needle-nosed pliers and check if the wire-clamp system is reliable. If the wire clamping system is not reliable, check the air supply line for leaks, whether the wire-clamp solenoid valve is reliable, and whether the cylinder for wire-clamp is dirty.

2.2.3. Turn on the mixed gas signal, check if the mixture gas pressure is between 0.4±0.05MPa, and adjust the mixture flow to 20~25 L/min.

2.2.4. The welding line must be accurately added to match the TCP point and a reasonable program running speed shall be selected.

2.2.5. When programming the welding procedure, the assignment range should be accurately determined, and repeated assignments are strictly prohibited.

2.2.6. After the programming is completed, the programmed welding procedure shall be simulated. The simulation adopts the single-step operation mode and the running speed of the program should be less than 75%.

2.2.7. After the simulation is completed, check whether the water circulation, air supply, and wire feeding system are normal. After confirmation above without error then enter into the automatic mode for automatic welding [3].

3. Analysis of tensile properties of plat butt welding

According to the GB/T228-2002 metal material tensile test method at room temperature, the CSS-44300 electronic testing machine is used for stretching, and the tensile test was carried out using a rectangular cross-section sample of 14.40×10.20 mm, and the stretching speed was 3 mm/min. Sample No. A is a welding wire for ESAB welding wire; sample No. B is a manual welding of the cohesive welding wire; and No. C is a automatic welding of the cohesive welding wir

IOP Conf. Series: Materials Science and Engineering **637** (2019) 012004 doi:10.1088/1757-899X/637/1/012004

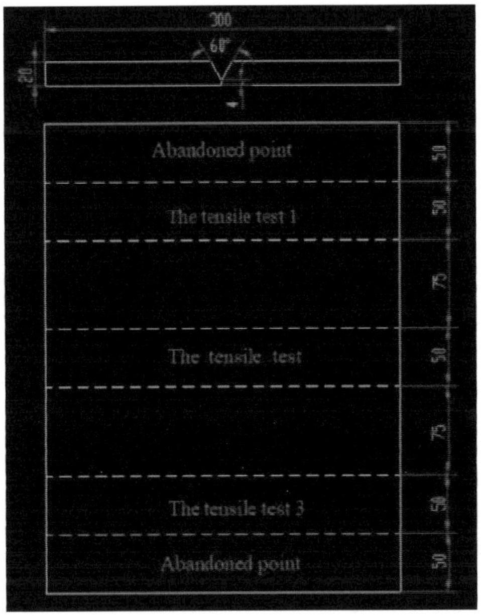

Figure 1. Schematic diagram of sampling position of plat butt welding tensile specimen

The welding sampling position is shown in Figure 1, the tensile test results are shown in Table 1, Figure 2, Figure 3, and Figure 4 show the sample of the tensile tests of samples A, B, and C:

Table 1. Tensile test results of plat butt welding

No.	Yield Strength MPa	Tensile strength MPa	Elongation δ 5%	Reduction of area Ψ%	Break position	Remarks
A1	——	503.5	25.8	55.8	Base metal	Yield is not obvious
A2	345.3	506.3	27.2	58.6	Base metal	
A3	——	506.9	27.2	59.7	Base metal	The yield is not obvious
Average value	345.3	505.6	26.7	58.0		
B1	342.5	499.5	26.8	60.5	Base metal	
B2	——	497.7	26	58.4	Base metal	The yield is not obvious
B3	342.1	499.5	27	58.0	Base metal	
Average value	342.3	498.9	26.6	59.0		
C1	356.6	493.2	30.0	58.0	Base metal	
C2	335.9	502.8	29.6	56.6	Base metal	
C3	344.6	489.9	30.0	57.3	Base metal	
Average value	345.7	495.3	29.9	57.3		

IOP Conf. Series: Materials Science and Engineering **637** (2019) 012004 doi:10.1088/1757-899X/637/1/012004

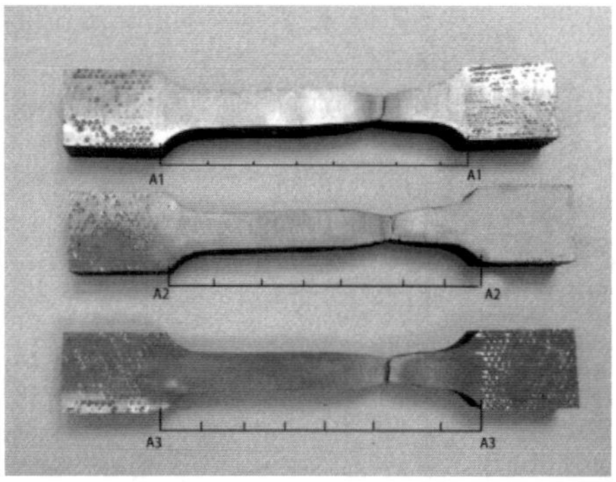

Figure 2. No. A tensile specimen

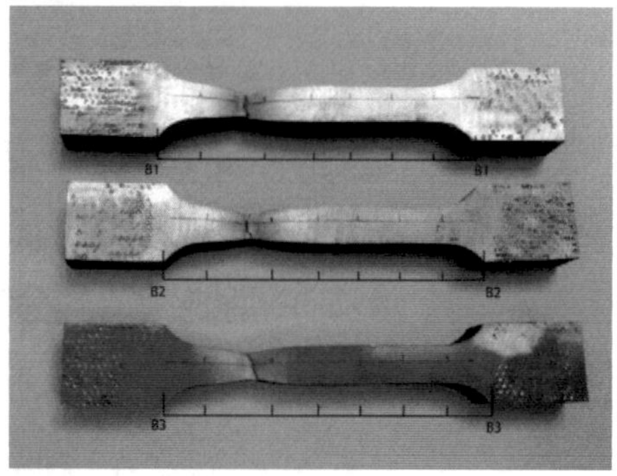

Figure 3. No. B tensile test specimen

Figure 4. No. C tensile test specimen

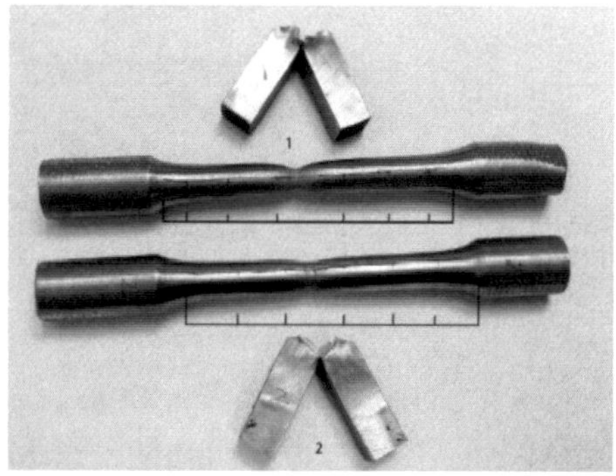

Figure 5. Specimens of deposited metal tensile and impact

The qualification standard for the normal temperature tensile test is that the tensile strength of the welded joint is not lower than the lower limit of the specified value of the tensile strength of the base metal, and the elongation of the weld metal is not less than 80% of the specified value of the parent metal [4]. The tensile test value mainly depends on the welding material and process, and is less affected by the welder's skill. Comparing the values in Fig. 2, Fig. 3, Fig. 4 and Table 1, it is not difficult to find that in the automatic welding, the tensile properties of the plate butt welding are in compliance with the requirements and is not too much different from the artificial welding.

4. Specimens of Deposited metal tensile and impact

According to the national standard GB/T228-2002 metal material tensile test method at room temperature, the CSS-44300 electronic testing machine is used for stretching, and the tensile test is carried out with Φ10×50mm sample, and the stretching speed is 3mm/min. The impact test uses a standard sample, sample No. 1 is a cohesive wire robot welding; sample No. 2 is a cohesive wire artificial welding. The sampling position of the deposited metal tensile and impact samples is shown in Figure 5:

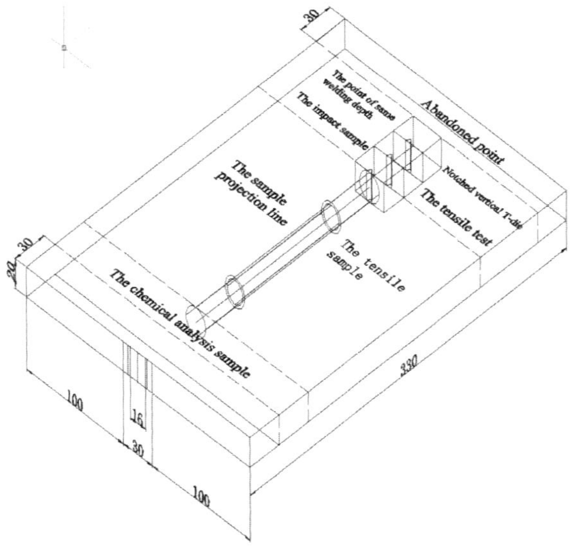

Figure 6. Schematic diagram of the sampling position of the deposited metal tensile and impact specimens

Results of the tensile test and impact test are shown in Table 2, and Figure 6 is a graph of specimen of the deposited metal tensile test and the impact test:

Table 2. tensile metal tensile test results

No.	Yield Strength MPa	Tensile strength MPa	Elongation δ_5%	Reduction of area Ψ%	Impact work J	Remarks
1	401.3	523.6	31.8	73.0	147	Unbroken
2	466.8	573.9	29.4	43.8	108	Unbroken

The test results of the two sets of tensile specimens were compared by Table 2, and the two groups were unbroken. Through the macroscopic analysis of the fracture morphology of Figure 6, both groups have obvious plastic deformation, and the fracture location is characterized by brittle fracture. According to the above test, the performance of the deposited metal of the automatic welding and the manual welding is not much different.

5. Weld microstructure analysis of T-joint

Figure 7 and Figure 8 show the macroscopic feature of the T-shaped weld seam. We use a cohesive welding wire and take two samples. The number of the cohesive welding wire samples is A and B, where A is robot welding and B is manual welding with an average penetration depth of 2.98 mm.

IOP Conf. Series: Materials Science and Engineering **637** (2019) 012004 doi:10.1088/1757-899X/637/1/012004

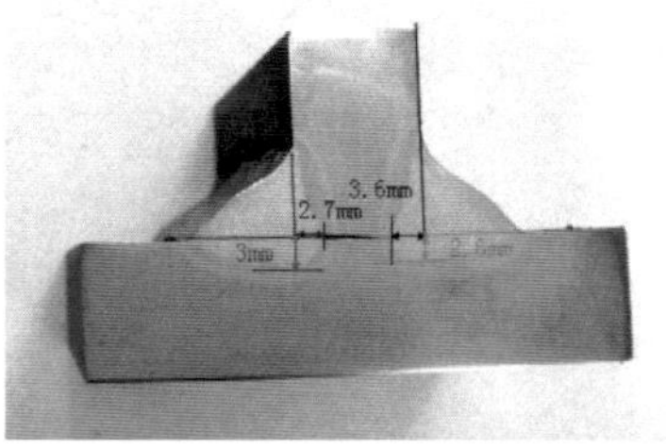

Figure 7. cohesive welding wire samples A

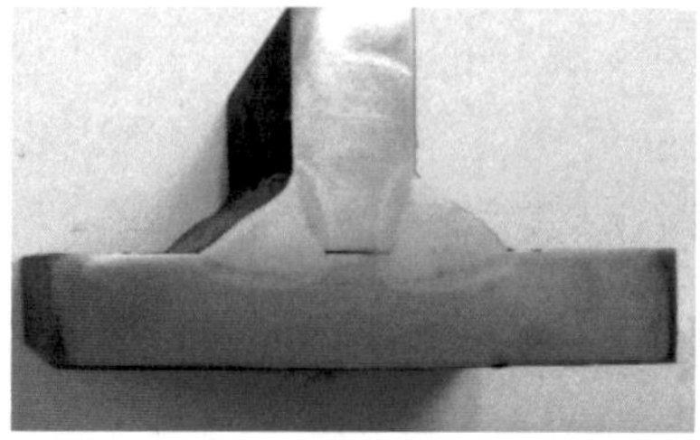

Figure 8. cohesive welding wire penetration sample B

The welding depth of the cohesive welding wire is 3mm, 2.7mm, 2.6mm, 3.6mm, and the form of weld seam is basically the same, which is the base metal, the incomplete crystallization zone, the complete crystallization zone (normalized zone), and the coarse grain zone, fusion zone, weld seam zone. Below is the microstructure of each sample.

A sample microstructure

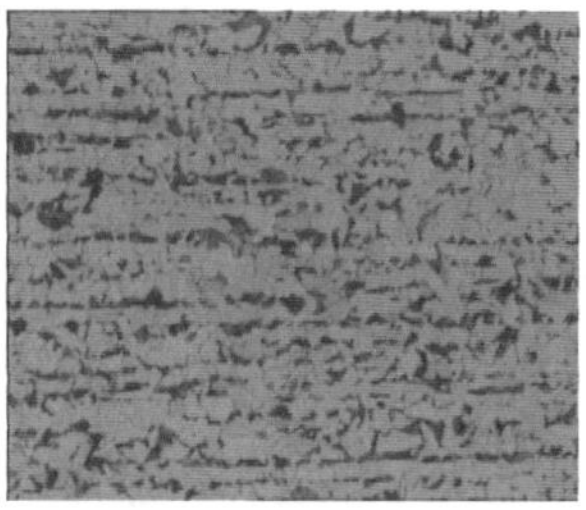

A base metal 200 times

A incomplete crystallization zone area 200 times

A complete crystallization area 200 times

A coarse grain area 200 times

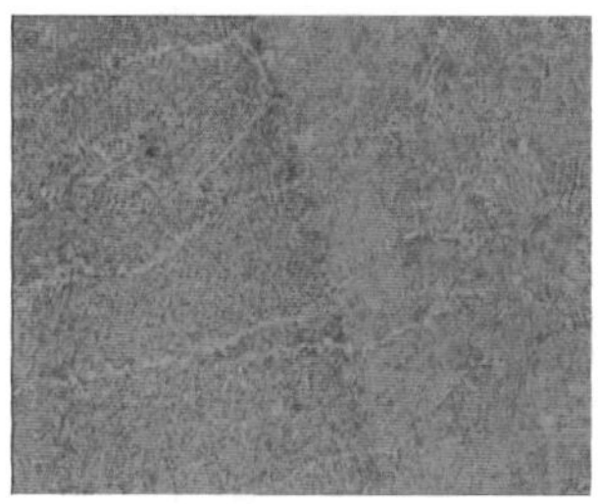

A fusion zone 200 times

A weld zone 200 times

B sample microstructure

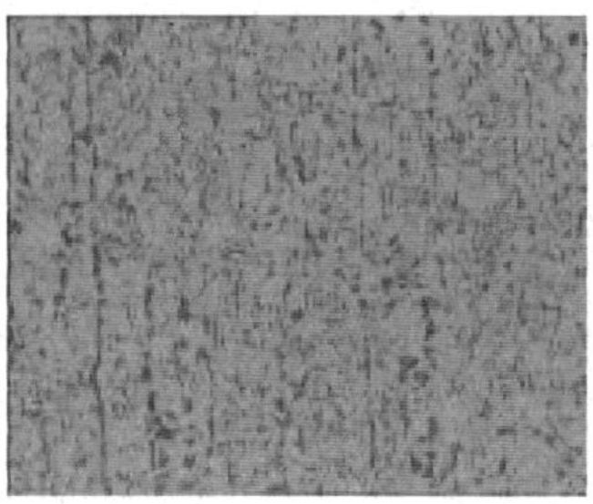

B base metal 200 times

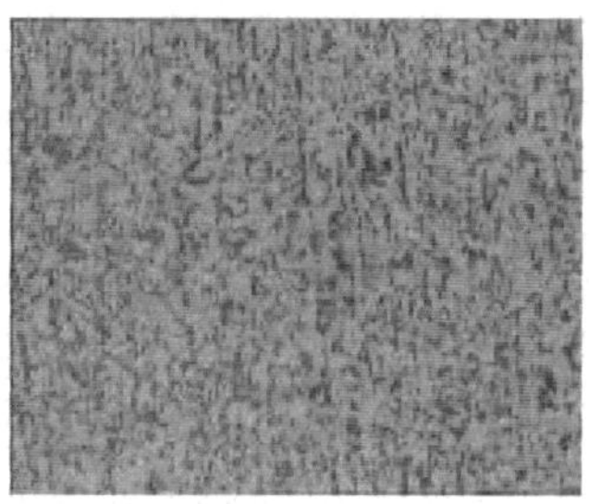

B incomplete crystallization

B complete crystallization

IOP Conf. Series: Materials Science and Engineering **637** (2019) 012004 doi:10.1088/1757-899X/637/1/012004

B coarse grain area 200 times

B fusion zone 200 times

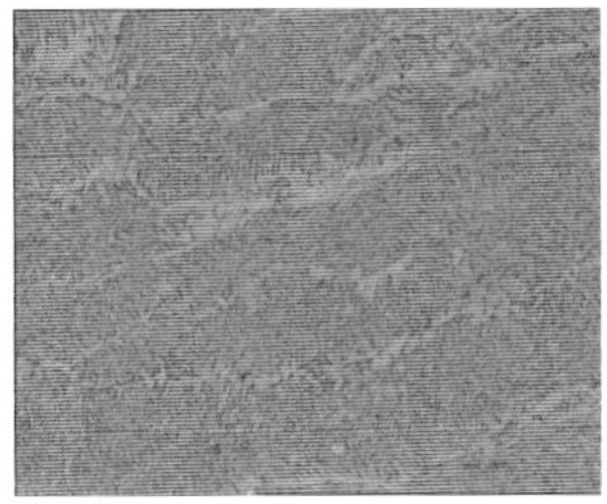

B weld zone 200 times

Analysis of sample microstructure characteristics:
The base material of the sample is ferrite and pearlite which is distributed in strips and is the same as the rolling direction of the sheet. In the incomplete crystallized area, a part of the structure undergoes a phase change which becomes fine ferrite and pearlite, and a part of it is the same as the base material. In the complete crystallization zone, the normalized zone, uniform and fine ferrite and pearlite, equivalent to the normalized structure during heat treatment [5]. In the coarse-grained zone, the superheated zone, the grain grows, there is Wei's structure, and the mechanical properties are poor. In the fusion zone, the upper part is the weld columnar crystal, and the lower part is the coarse crystal zone, which has a great influence on the welded joint and shaping. In the weld zone, columnar crystals.
Results of sample analysis:
It can be known from the hardness curve that the hardness of the samples A and B remained unchanged. In addition, it can be seen that the pearlite content in the area is increased compared with the base material, and the hardness of the weld bead is higher than that of the base material. On the one hand, it is the main reason of high carbon content. On the other hand, observing the structure of the weld beam we can know that there are columnar crystals inside, which is the main reason of the hardness.

6. The conclusion

The author compares the welding performance of automatic welding with manual welding by tensile test, impact test and microstructure analysis of welded plates, and it is worthwhile to determine the welding performance of automatic welding. However, if there is no suitable structural form, mature welding process, stable point quality, and excellent personnel quality, it cannot obtain the ideal welding quality even with the advanced welding technology. In the actual plate welding, a human-machine combination method can be adopted, which can control the welding cost and ensure the welding quality, and has certain engineering value.

References:

[1] Qing Weifu, Luo Bin.Application of Arc Sensing Technology in Weld Seam Tracking,J.*Metal Processing(Hot Working)*,(2014)40-41.

[2] Le Jian.Research on tracking rectangular welds based on rotating arc sensing welding robot,D. *Chinese excellent master's thesis full-text database*, 2018.

[3] Zhang Shengnan, Zhao Wenbin.Practice Plan of " Operation of Welding Robot" Course in Higher Vocational Colleges, J. *Journal of Mechanical Research and Application*, (2017)197-199.

IOP Conf. Series: Materials Science and Engineering **637** (2019) 012004 doi:10.1088/1757-899X/637/1/012004

[4] Li Yali, Chen Dechao.Characteristics Analysis of Charpy impact test based on -196 °C ultra-low temperature metal materials , J. *Industrial Construction*, (2018)156-158.
[5] Information on http://www.doc88.com/p-7854217685334.html

The Development Of Sensing Architecture For Inferior Alveolar Nerve Block Clinical Simulator Kit

S Z Zainudin[1], M H Mohd Ramli[1], T I B Tengku Jamaluddin[2], H Abdul Wahab[2], A Husin[2] and S A Abdullah[1]

[1]Faculty of Mechanical Engineering, Universiti Teknologi MARA, Malaysia
[2]Centre of Oral & Maxillofacial Surgery Studies, Faculty of Dentistry, Universiti Teknologi MARA, Malaysia
Email: syafiqzulhady1994@gmail.com

Abstract. This paper presents the development of sensing architecture for Inferior Alveolar Nerve Block (IANB) clinical simulator kit. The ultimate objective of this project is to emulate the environment of anaesthesia administration on an IANB which considers the integration of 3D printing technology and internet of things (IoT) concept. The 3D model of a lower jaw is obtained through Digital Imaging and Communications in Medicine (DICOM) data via dental Cone Beam Computed Tomography (CBCT) on a patient. The obtained model is further refined to facilitate the integration of the electronic circuit by utilization of a commercial computer-aided design (CAD) software package. The printed model is then integrated with a sensing mechanism to provide real-time data feedback during simulation of local anaesthesia administration training. From the validation results, it shows that the prototype able to give real-time data feedback and display the results on the user interface.

1. Introduction

Local anaesthesia is an essential part of dental treatment and is formally taught in the dental curriculum. It is compulsory for dental students to study and understand the relevant concepts of anatomy that is crucial for the administration of local anesthesia [1]. The inferior alveolar nerve block (IANB) is one of the most frequent local anaesthesia techniques (injection) used in dental procedures [2]. However, this technique is reported to have a high failure rate among dental students and interns [3]. Thus, it is imperative for dental students to have proper practical training in order to reduce mistakes and ensure that they master local anaesthesia administration skills prior to treating real patients in their clinical years. Currently, the common practice in pre-clinical training is through student-to-student administration of local anaesthesia in most dental schools worldwide [4]. This method is proven to be effective in mastering the anaesthesia administration technique. The above-mentioned conventional teaching method is also exposed to potential risks such as when it is injected at the wrong position, i.e, beneath the mandibular foramen or too close to the anterior border of the ramus which could result in physical complications [5]. For instance, if the needle is positioned too superiorly or posteriorly, the anaesthesia may pass the parotid gland and anaesthetize the Facial Nerve or Cranial Nerve VII that may cause transient facial paralysis. Symptoms of this temporary loss will affect the motor functions of the muscles of facial expression which includes the inability to close the eyelid and drooping of the labial commissure on the affected side for a few hours [6]. In another example, if the needle is placed too medially, the medial pterygoid muscle (a muscle of mastication) can be injected, resulting in trismus,

IOP Conf. Series: Materials Science and Engineering **637** (2019) 012005 doi:10.1088/1757-899X/637/1/012005

which is the inability to open the mouth fully [7]. Furthermore, this conventional teaching method is also prone to be dangerous if conducted in the event when the students are distracted or psychologically unfocused.

It is strongly believed that modern teaching and assessment methods could address the above problems. A simulator kit with a simulation-based training will provide students with skills and competencies in handling anaesthesia administration [8]. To effectively reduce and alleviate the physical complications and further ensure that dental students master the correct anaesthesia administration technique, we propose an IANB clinical simulator kit. The simulator kit facilitates the students with repetitive practice and data feedback as an educational platform to achieve precision in pre-clinical administering local anaesthesia training prior to treating real patients.

Historically, since the 1930s, simulation-based assessment has been deployed in the health care field, which much of it derived from military and aviation industries [9]. The simulation may exist in different forms such as virtual patients, interactive mannequin simulators and screen-based (computer) simulation that provide users with knowledge and experience that mimics real-life application [10]. Such simulator features can provide training and assessment to replace conventional teaching methods in the dental curriculum. Augmented reality (AR) and virtual reality (VR) are among recent technology that is used in developing the simulation for the IANB training [11-12]. Apart from that, the physical type of simulator has also been developed and used for IANB administration training. The development of these simulator has progressed positively since two decades ago to further improve the sensing mechanism and sensation during injection training [13]. As advancements in electronics and mechanical aspect in manufacturing namely additive manufacturing, numerous applications such as measurements and complex models of the human part become viable, therefore, an alternative sensing mechanism is proposed in the direction of developing a physical simulator kit for administering of local anaesthesia training skills.

2. 3D Modeling of Human Lower Jaw

This section discusses the first requirement in realizing the simulator prototype of the lower jaw (mandible) model that could mimic the human lower jaw. For this purpose,

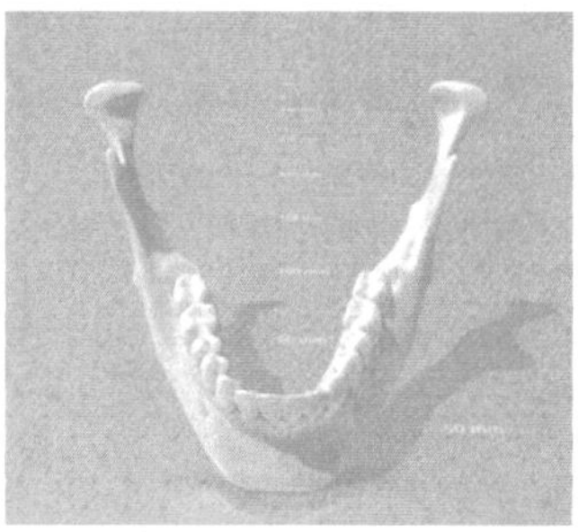

(a) Front view of the 3D model

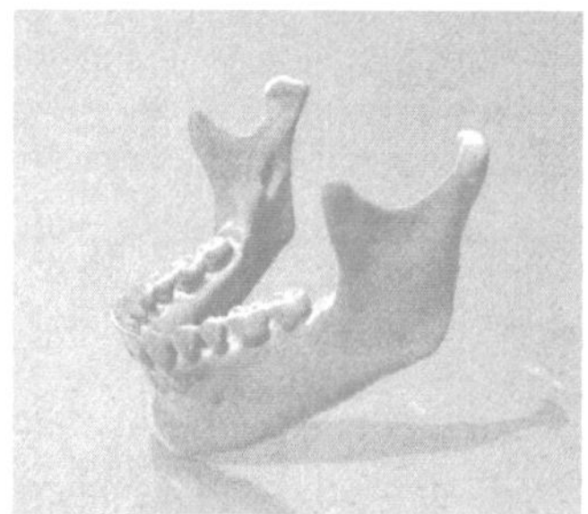

(b) Side view of the 3D model

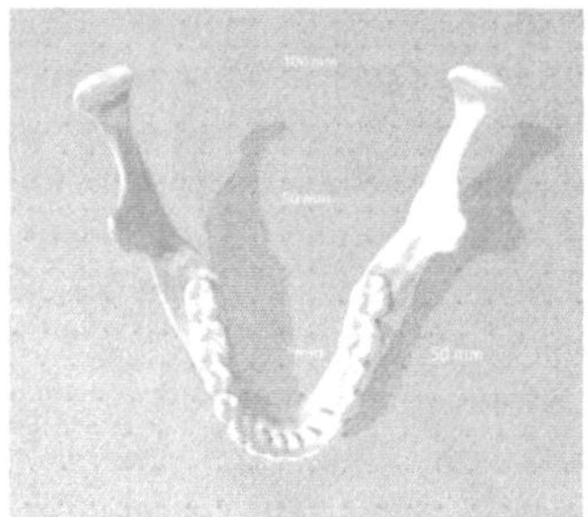

(c) Top view of the 3D model

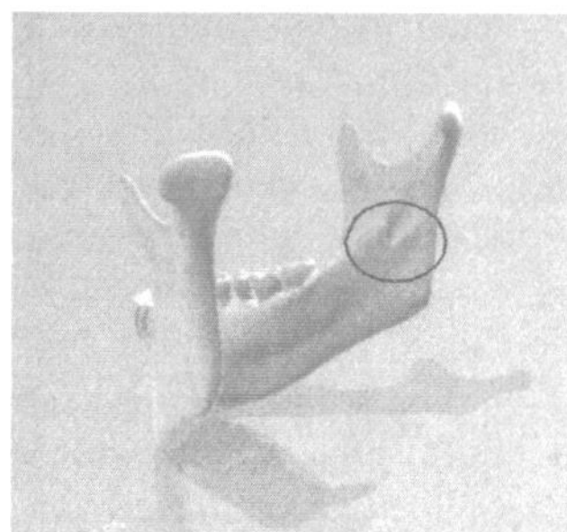

(d) Target location of IANB administration at the red circle

Figure 1. Illustration of (a) Front view of the 3D model, (b) Side view of the 3D model, (c) Top view of the 3D model, (d) Target location of IANB administration at the red circle a Digital Imaging and Communication in Medicine (DICOM) data was used to generate the 3D model of the lower jaw including nerve pathways and bone which was obtained via Cone-Beam Computed Tomography (CBCT) scan on a patient. The CBCT image files were converted and stored in DICOM, and then, extracted and imported to Simpleware® software package for rendering process of a 3D model of the lower jaw. The Simpleware® software is able to open DICOM files and subsequently, convert the files to stereolithography (STL) files which are the crucial step of producing the 3D model. However, further refinement of the 3D model is required to facilitate the integration of the electronics, commercial CAD software was utilized. The refinement process in this sense included structure manipulation to create passageways and slots for the wiring harness, micro-controller, and sensors.

The 3D model was built by using fusing deposition modeling (FDM) which is categorized in the additive manufacturing process. The built prototype is integrated with electronic circuits and sensors to give feedback for the training purpose. Figure 1 shows the completed 3D model in perspective views.

3. Integration of Electronics

This section discusses the integration of electronics and circuits on the printed prototype to provide a feedback system for the IANB simulator kit. Core electronics elements include Arduino Pro Mini and capacitive sensor. 18mm x 33mm Arduino Pro Mini in dimension is considered small and has the capacitive features onboard. This microcontroller is used to process data from sensors and communicates with the user through the serial monitor via a USB connection. Capacitive sensors are used as sensing mechanisms on the prototype. 10 capacitive sensors are embedded on target locations of the IANB administration (mandibular foramen) as representative of the target area that will be assessed during the injection. Figure 2 shows the illustration of the prototype with the integration of electronics and circuit.

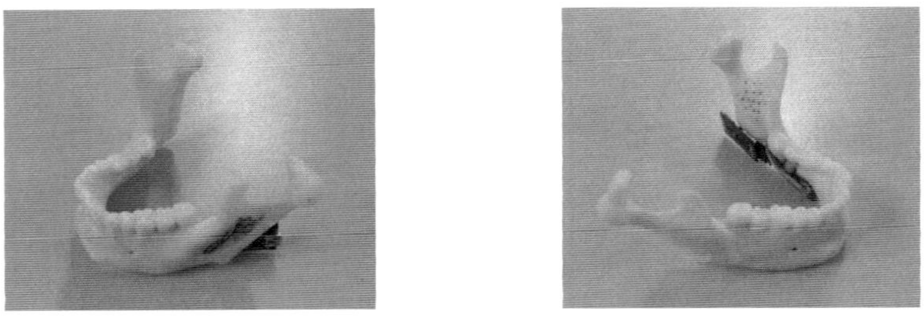

Figure 2. Illustration of the prototype with the integration of electronics and circuit

The prototype is programmed to give feedback and display its result to the users. The operational flow of the prototype is shown in Figure 3.

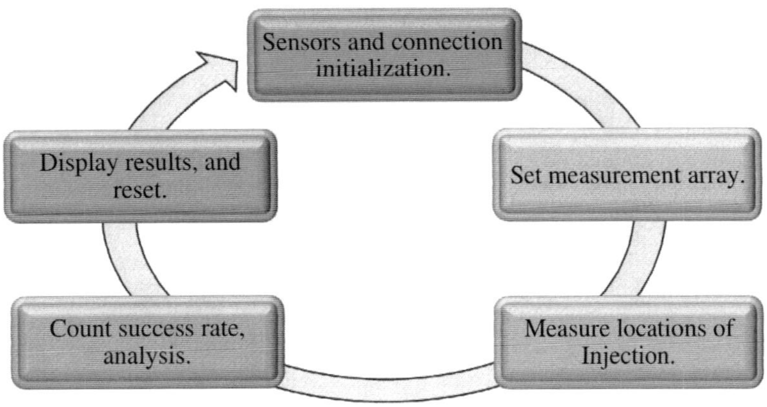

Figure 3. Logic of operation process flow for IANB simulator kit

IOP Conf. Series: Materials Science and Engineering **637** (2019) 012005 doi:10.1088/1757-899X/637/1/012005

4. Prototype Testing and Validation

This section discusses prototype testing by replicating the real IANB administration technique. A syringe is partially modified by winding the copper wire around it to activate the capacitive sensor when an injection is performed. The copper winding will allow electric current flow from the user's hand to the capacitive sensor on the simulator kit. Figure 4 depicts the syringe that was utilized in the trial session.

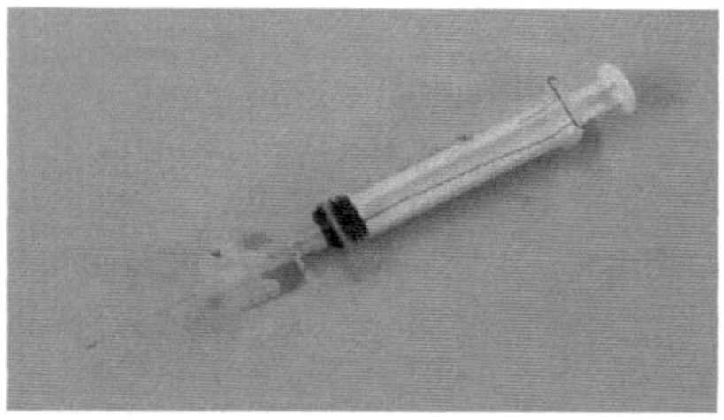

Figure 4. Illustration of copper wire is wound up around syringe

The functionality of the prototype was tested and evaluated by dental students. They injected the prototype at several locations to view the results of the trial session, as shown in Figure 5.

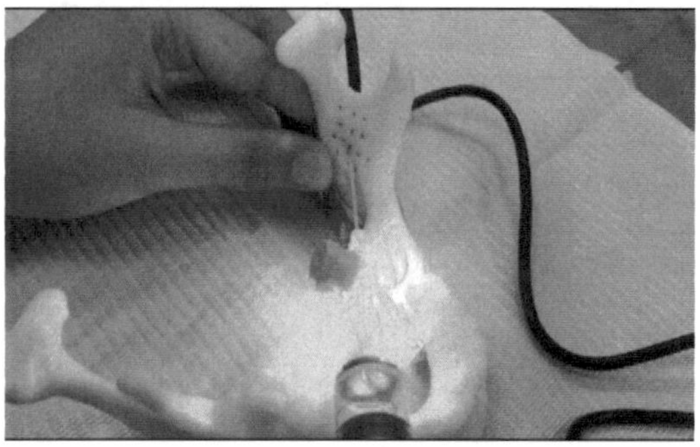

(a) Accurate position during trial session

(b) Acceptable position during trial session

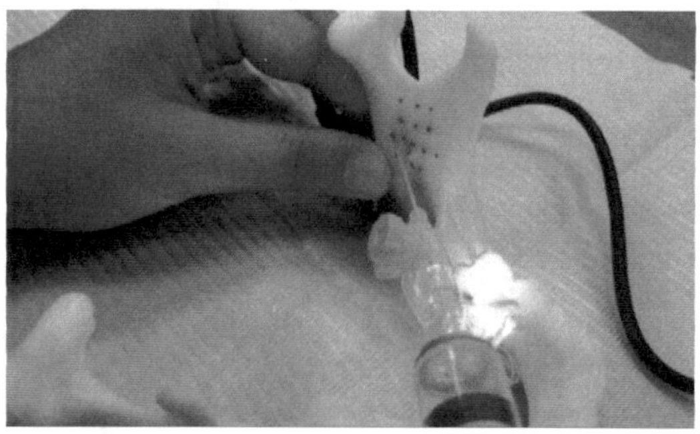

(c) Incorrect position during trial session

Figure 5. Illustration of (a) Accurate position during trial session, (b) Acceptable position during trial session, (c) Incorrect position during trial session

Colour code is used to differentiate the assessment results, i.e., LED light. The colour indication is working fine as it turns green signifying accurate position, blue for an acceptable position and red for incorrect position. The physical indicator (LED) eases students to be aware of their injection accuracy status. Apart from that, the serial monitor displays the result as well to the users in the digital environment. Figure 6 shows the result of the injection at different locations displayed on the serial monitor.

Figure 6. Illustration of the serial monitor depiction of an accurate and incorrect injection on the prototype

The students gave positive feedback on the trial session as the prototype was capable to provide users with real-time feedback of the accuracy of the user's injection. The assessment of the injection was clearly indicated by different LED colours. Apart from that, the result is shown in the serial monitor with a description of the injection which gives detailed information to the users. The sensing mechanism shows good results and the prototype performed reasonably well upon the prescribed task.

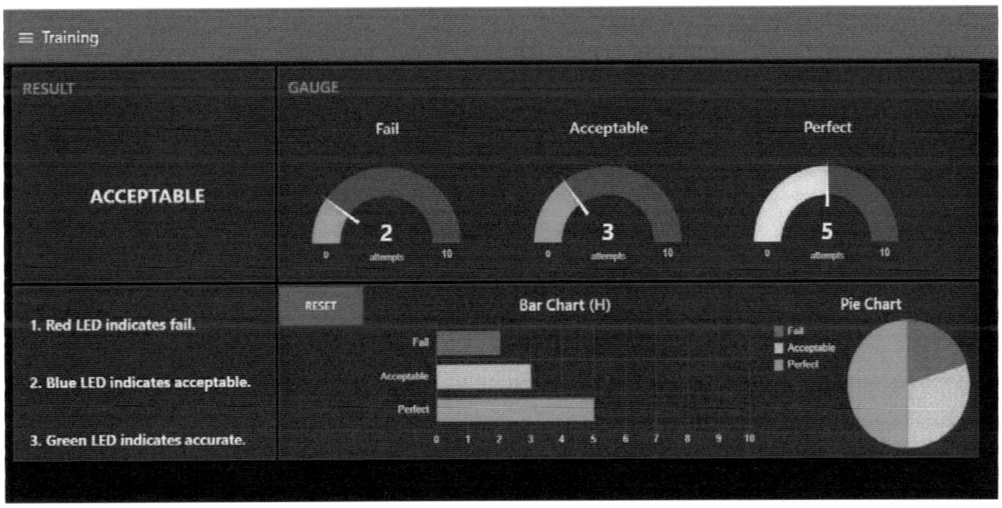

Figure 7. Illustration of the user interface with data analysis

The user interface display the results with data analysis to the users as shown in Figure 7. The user interface can be accessed by the users during the training session to monitor the performance of the user with data analytic in term of successful attempt with graphical charts. Real-time data feedback from the simulator is displayed in the user interface with the availability of internet connection.

5. Further Direction for the Prototype

This section discusses improvements that can be made to the prototype in order to create a more comprehensive IANB simulator kit for IANB administration training. The sensing mechanism will be encapsulated with a composite material that mimics the properties of human gums. By encapsulating the prototype, the training becomes more challenging as the users do not see the landmarks of the embedded sensors.

The life-like sensation of the IANB administration is an important criterion to be considered for IANB training. Therefore, the prototype will be inserted into dental mannequin heads easily available at the dental faculty such as shown in Figure 8.

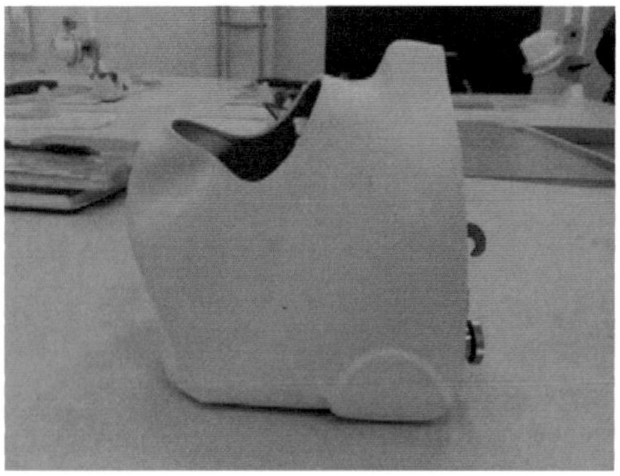

(a) Side view of the removable head mannequin

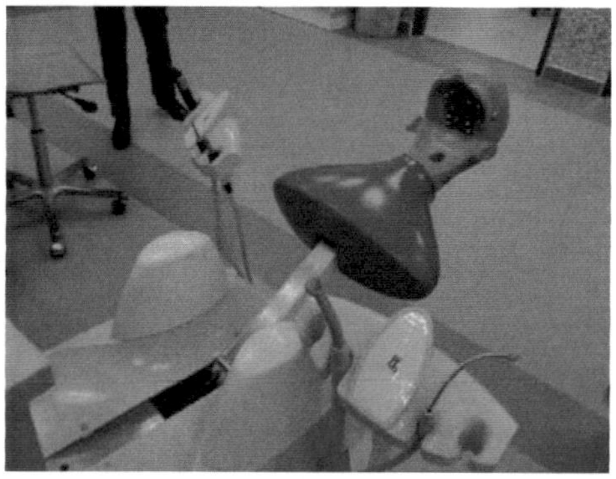

(b) Upper body mannequin with the removable head

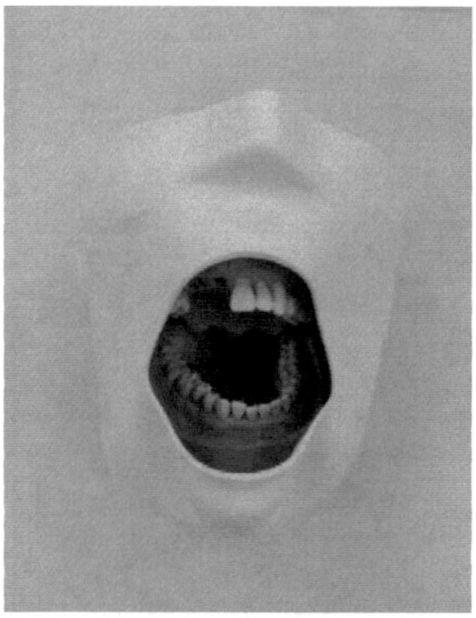

(c) Top view of the removable head mannequin

Figure 8. Illustration of (a) Side view of a removable head mannequin, (b) Upper body mannequin with removable head, (c) Top view of the removable head mannequin

The advancement of technology and in particular IoT utilized as a smart assessment system for the simulator kit. A graphical user interface (GUI) will be used to display and show training results for each student by allocating their data in a database system. Each student can log in to the system before training commences. Therefore, lecturers can assess their students based on records and trends of the training. Systematic training method can be applied to replace the conventional training method with a complete online assessment system.

6. Conclusion

Current stage development of the prototype has met with the requirements of practitioners in the dental profession as well as within the dental education environment. The prototype will be able to act as the primary learning platform for the administration of local anaesthesia in dental schools before continuing with student-to-student injection method. The current stage of the prototype was able to provide profound physical and digital environment feedback which is considered significant information for the users. However, there are several concepts that are under consideration to further enhance the simulator kit. One of the ongoing efforts is to encapsulate the simulator kit prototype with suitable composite material to resemble the human gum. We aspire that the durability of the composite material will not leave definite marks on the model and provide a real tactile sensation of the oral environment for the user when performing the IANB injection. Thus, each student can train repeatedly on the model due to its durability. The lower jaw model might be attached to a human mannequin to provide a life-like training on a real patient instead of a stand-alone simulator as per the conventional teaching method. Apart from that, data stored from training sessions may be used for precise assessment and record-keeping purposes to achieve an effective teaching and training system. Overall the IANB simulator kit will provide a profound teaching and learning experience for both students and educators as well as being able to improve team resource management in the dental clinical education environment.

Acknowledgment
This work was supported in part by the UiTM Internal Grants MITRA PERDANA (600-IRMI/DANA 5/3/MITRA (P) (002/2018-1)).

References

[1] Hossaini M 2011. Teaching local anesthesia in dental schools: opinions about the student-to-student administration model. *J Dent Educ.* **75**(9):1263–9.
[2] Sabari H, Kumar S, Andrew A, Anandan H 2018. Comparison of Anesthetic Efficacy of IANB -Classical and Gow-Gates Technique during Surgical Removal of Impacted Mandibular Third Molar. *Ann Int Med Dent Res.* **4**(2):69–72.
[3] Alhindi M, Rashed B, Alotaibi N 2016. Failure rate of inferior alveolar nerve block among dental students and interns. **37**(1):84–9.
[4] Chandrasekaran B, Cugati N, Kumaresan R 2014. Dental students' perception and anxiety levels during their first local anesthetic injection. *Malaysian J Med Sci.* **21**(6):44–50.
[5] Canavan D 2018. Delivering painless and effective inferior alveolar nerve block anaesthesia Delivering painless and effective inferior alveolar nerve block anaesthesia. *Irish Dent Assoc.* **60**:230–2.
[6] V.R. S, M.S. T 2014. Facial nerve paralysis after anaesthetic usage- A review. *J Pharm Sci Res.* **6**(9):308–9.
[7] Monisha N, Ganapathy D, Sheeba PS, Kanniappan N 2018. Trismus : A review. 12(2):10–3.
[8] Cook DA, Andersen DK, Combes JR, Feldman DL, Sachdeva AK 2018. The value proposition of simulation-based education. *Surgery.* **163**(4):944–9.
[9] Bradley P 2006. The history of simulation in medical education and possible future directions. *Med Educ.* **40**(3):254–62.
[10] Burden A, Pukenas EW 2018. Use of Simulation in Performance Improvement. *Anesthesiol Clin.* **36**(1):63–74.
[11] Won Y-J, Kang S-H 2017. Application of augmented reality for inferior alveolar nerve block anesthesia: A technical note. *J Dent Anesth Pain Med.* **17**(2):129–34.
[12] Tori R 2017. Virtual Reality simulator for dental anesthesia training in the inferior alveolar nerve block. **25**(4):357–66.
[13] K. Makoto 2017. The development of a simulator from 1992 to 2016 for inferior alveolar nerve block injection and skill education for delivering local injections of dental anesthetics. Dent J Heal Sci Univ Hokkaido.

IOP Conf. Series: Materials Science and Engineering **637** (2019) 012006 doi:10.1088/1757-899X/637/1/012006

Redesigning of Agarwood Extracting Machine Applying DFMA Principle

M S Salim[1,a], M A Lajis[1,b], Z C Ros[2,c], A Nawawi[3,d], S Shamsudin[1,e], N K Yusuf[1,f]

[1]Sustainable Manufacturing and Recycling Technology, Advanced Manufacturing and Materials Center (SMART-AMMC), Universiti Tun Hussein Onn, Malaysia
[2]Center for Diploma Studies, Universiti Tun Hussein Onn, Malaysia
[3]Faculty of Engineering Technology, Universiti Tun Hussein Onn, Malaysia
[a]syazwansalim4@gmail.com; [b]amri@uthm.edu.my; [c]ziana@uthm.edu.my; [d]azle@uthm.edu.my; [e]shazarel@uthm.edu.my; [f]nurkamilah@uthm.edu.my

Abstract. This paper presents the comparative study of design efficiency of two different designs of agarwood extracting oil machines by performing using the Design for Manufacturing and Assembly (DFMA) method. The DFMA method was used as it is a well-established technique for improving the efficiency of the product leading to minimizing production costs. It also shortens product development time by reducing the number of components in a product. The study used two different designs of agarwood extracting oil machines as a case study. The result shows that the percentage of the design efficiency of existing design model 1 is 9.25%, whereas it is 15% of redesign. Thus, the redesign model 2 is much better as compared to existing design model 1 in terms of its assembly operation and design efficiency. Therefore, the redesign model 2 is greener than existing design model 1. Therefore, the application of the DFMA method to enhance the development of a agarwood extracting product has been proven to be highly useful in the design work.

1. Introduction

This project includes a study of manufacture and assembly costs estimation of an agarwood system which is used to extract agarwood oil. In this project Agarwood extracting machine was designed by applying DFMA principle. Design for Manufacturing (DFM) is the design for ease of manufacturing from collection of parts that will form the product and Design for Assembly (DFA) is the design for ease of assembly. The aim of DFMA is to reduce manufacturing and assembly costs, and to quantify improvements [1]. The practice of applying DFMA is to identify, determine and eliminate wasteful or inefficient product design. DFMA is also used as a benchmarking tool to review product competitors, and as a tool to aid in cost negotiations suppliers [2]. The final assembly cost of the designed product was compared to that of the existing design available in the market and the results are discussed. Hydro-distillation used in the production and extraction of essential oils. This is often the best and infrequently the most cost-effective distillation method. Hydro-distillation looks to figure best for powder and materials terribly tough as root, wood, or bean. The most advantage of this technique is that the steam is a smaller amount used, shorter interval and better oil yields [3].

2. Methodology

2.1. Selection of Product for the Case Study

IOP Conf. Series: Materials Science and Engineering **637** (2019) 012006 doi:10.1088/1757-899X/637/1/012006

In return for research objectives, two of Agarwood extraction machine based on hydro distillation method has been selected which are the original design and the redesign of the machine as shown in Figures 2 and 3 respectively. The detail drawing of each component was reproduced with use of Solidworks modelling software. Design for Assembly (DFA) and Design for Manufacturing (DFM) was computed using Boothroyd Dewhurts DFMA method [4].

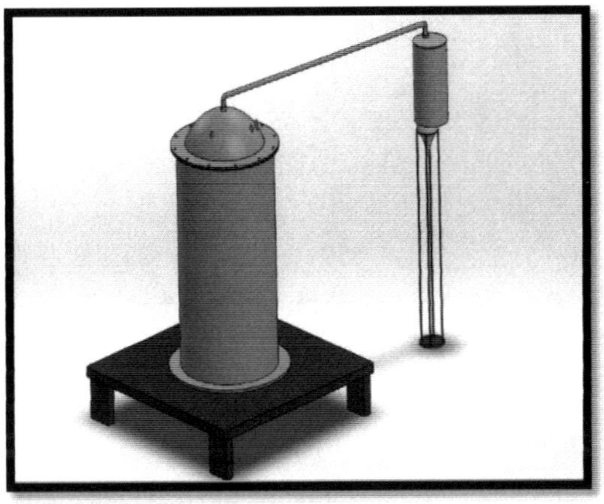

Figure 1. Complete design of original product

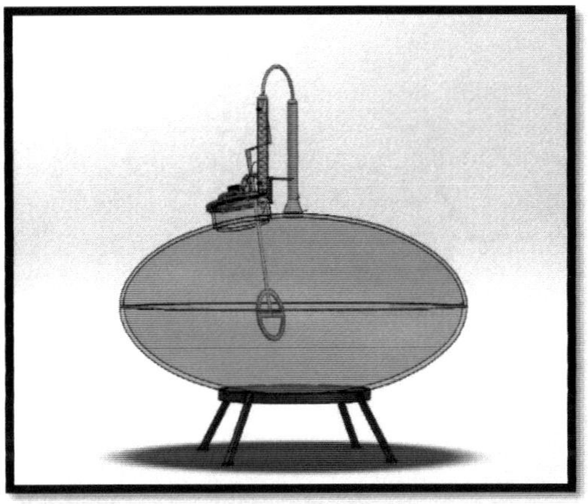

Figure 2. Complete design of redesign

2.2. Overall Steps to Compute Design Efficiency

The design efficiency analysis is carried out on two factors, namely i) consideration of the possibility to remove parts or to join other parts of the assembly, and ii) estimates of estimated time taken to grasp, manipulate, and insert that part. The following procedure was adopted to apply the DFMA in the agarwood extracting machine. The calculation of design efficiency, DE is based on the Equation (1) below;

$$DE = {(3NM)}/{TM} \quad (1)$$

where *NM* is the theoretical minimum number of parts and *TM* is the total manual assembly time (seconds).

3. Result and discussion

3.1. Design for Manual Assembly Worksheet for Existing Design (Model 1)

Sections should be numbered with a dot following the number and then separated by a single space

Table 1. Design efficiency of the original product or existing design.

0	C1	C2	C3	C4	C5	C6	C7	C8	C9

IOP Conf. Series: Materials Science and Engineering **637** (2019) 012006 doi:10.1088/1757-899X/637/1/012006

Name of part	Part ID	No of operations carried out consecutively	Two-digit manual handling code	Manual handling time per part	Two digit manual insertion code	Manual insertion time per part	Operation time C2(C4+C6)	Operation cost,(RM) x (7)	Estimation for theoretical minimum parts
Table leg	1	4	00	1.13	96	12	52.52	1.313	0
Table base	2	1	94	3	00	1.5	4.50	0.113	0
Base boiler	3	1	90	2	96	12	14.00	0.350	0
Boiler	4	1	96	4	20	5.5	9.50	0.238	1
Hex bolt M5x0.8x25	5	8	00	1.13	00	1.5	21.04	0.526	0
Hex bolt M16x2x65	6	12	00	1.13	00	1.5	31.56	0.789	0
Hex flange nut	7	12	02	1.88	00	1.5	40.56	1.014	0
Lid boiler	8	1	92	2	10	4	6.00	0.150	1
Lid mouth	9	1	00	1.13	96	12	13.13	0.328	1
Pipe connector	10	1	90	2	08	6.5	8.50	0.213	1
Coil condenser	11	1	92	2	18	9	11.00	0.275	1
Coil condenser tank	12	1	92	2	18	9	11.00	0.275	1
Coil condenser lid	13	1	00	1.13	06	5.5	6.63	0.166	0
Beaker	14	1	00	1.13	00	1.5	2.63	0.066	1
Beaker ring	15	1	01	1.43	00	1.5	2.93	0.073	1
Beaker leg	16	4	00	1.13	96	12	52.52	1.313	1
Beaker base	17	1	00	1.13	01	2.5	3.63	0.090	0
Design Efficiency= 3NM/TM, 3(9)/291.65 = 0.0925							**TM**	**CM**	**NM**
							291.65	**7.292**	**9**

3.2. Design for Manual Assembly Worksheet for Redesign (Model 2)

Table 2. The design efficiency of the redesign product.

0	C1	C2	C3	C4	C5	C6	C7	C8	C9

IOP Conf. Series: Materials Science and Engineering **637** (2019) 012006 doi:10.1088/1757-899X/637/1/012006

Name of part	Part ID	No of operations carried out consecutively	Two-digit manual handling code	Manual handling time per part	Two digit manual insertion code	Manual insertion time per part	Operation time C2(C4+C6)	Operation cost,(RM) x (7)	Estimation for theoretical minimum parts
Cylindrical funnel	1	1	00	1.13	96	12	13.13	0.250	1
Upper boiler	2	1	94	3	00	1.5	4.50	0.086	1
Cone funnel	3	1	00	1.13	96	12	13.13	0.250	0
Lower boiler	4	1	94	3	06	5.5	8.50	0.162	1
Pipe connector	5	1	92	2	02	2.5	4.50	0.086	1
Lid	6	1	90	2	00	1.5	3.50	0.067	1
Stirrer rod	7	1	90	2	07	6.5	8.50	0.162	1
Lid holder	8	2	00	1.13	96	12	13.13	0.250	0
Hex bolt M6x1.0x35	9	1	00	1.13	39	8	9.13	0.173	1
Hex flange nut M6	10	1	00	1.13	39	8	9.13	0.173	0
Motor housing	11	1	10	1.5	96	12	13.50	0.256	1
40 rpm motor	12	1	05	1.84	01	2.5	4.34	0.082	1
Straight bevel gear	13	1	05	1.84	02	2.5	4.34	0.082	1
Base boiler	14	1	90	2	06	5.5	7.50	0.143	0
Leg	15	4	90	2	96	12	14.00	0.266	0
Connector	16	1	00	1.13	96	12	13.13	0.250	0
Beaker hook	17	1	00	1.13	00	1.5	2.63	0.050	1
Beaker	18	1	00	1.13	00	1.5	2.63	0.050	1
Hex bolt M10x1.5x55	19	9	00	1.13	10	4	46.17	0.087	0
Hex flange nut M10	20	9	00	1.13	11	5	55.17	1.048	0
Straight bevel pinion	21	1	05	1.84	07	6.5	8.34	0.158	1
Gear stopper	22	1	44	5.6	07	6.5	12.10	0.230	0

Stirrer holder	23	1	00	1.13	03	3.5	4.63	0.088	0
Coil condenser	24	1	92	2	08	6.5	8.50	0.162	1
Condenser tank	25	1	92	2	02	2.5	4.50	0.086	1
Condenser lid	26	1	00	1.13	96	12	13.13	0.250	0
Design Efficiency= 3NM/TM, 3(15)/301.76 = 0.150 (15 %)							**TM**	**CM**	**NM**
							301.76	**4.947**	**15**

The design efficiency of two different design of the product was analyzed using Boothroyd Dewhurst DFMA methods for manual assembly. Tables 1 and 2 show the analysis results of the original product and the redesign product. The total assembly time for original product was 291.65s with 9 theoretical minimum part counts which leads to the design efficiency of 9.25%. The total assembly time for the redesign was 301.76s with 15 theoretical minimum part count which was leading to the design efficiency of 15%. Based on the results, the total parts for the existing design was 17 while for the redesign is 26 as shown in Table 3 below.

Table 3. Comparison between the existing design and redesign.

Item	Existing design	Redesign
Percentage of design efficiency	9.25%	15%
Number of parts	17	26
Total assembly time	291.65s	301.76s
Total cost	RM 38,568.84	RM 36,137.71

The percentage of design efficiency of existing design was 9.25% and for the redesign was 15%. The DFMA analysis results shows the redesign is more superior compare to existing design because the total cost for redesign were cheaper than original design because of the suitable material has been chosen for each part but the total assembly time for redesign was 301.76s which was more than original product which was 291.65s. This is because the total number of parts for the redesign was greater than original product which was 26 parts all of them while the original products just 17 parts.

4. Conclusion

The number of components and assembly methods is an important factor affecting the design efficiency of a product. In this study, the total number of components for original product components was increase from 17 components to 26 by using DFA analysis. Total assembly time increase from 291.65s to 301.76s that's why these projects only focus on the DFM which the total cost was reduced from RM 38,568.84 to RM 36,137.71. It's obviously cut the cost which was about RM 2423.31.79 or 16%. The reduction of the total cost is contributed by changing the material form stainless steel to aluminium based on DFMA analysis result. The reduction of cost can cut the cost and can increase the profit.

5. References

[1] Chowdary B V and Harris A 2009 Integration of DFMA and DFE for Development of a Product Concept : A Case Study. *Seventh LACCEI Latin American and Caribbean Conference for Engineering and Technology (LACCEI'2009) Energy and Technology for the Americas: Education, Innovation, Technology and Practice (June)* 1–8

[2] Batalha G F 2012 Concurrent engineering and DFMA/DFX in the development of automotive components *Open Access Library* **6** 12 30–51.

[3] Atikah N, Yusoff M, Tajuddin S N, Hisyam A, Adila N and Omar M 2015 Agarwood Essential Oil: Study on Optimum Parameter and Chemical Compounds of Hydrodistillation Extraction *Journal of Applied Science and Agriculture J. Appl. Sci. & Agric* **10**(105) p 1–5.
[4] Boothroyd G, Dewhurst P and Knight W 2002 Product Design for Manufacture and Assembly 2nd ed (New York: Marcel Dekker)

Acknowledgments
The authors would like to express their profound gratitude and deepest appreciation to the Universiti Tun Hussein Onn Malaysia (UTHM), Malaysia, for funding this project through the TIER-1 research grant scheme of vot number H-217. The authors would also like to express their special thanks to DFMA laboratory (MKIE) staff under Faculty of Mechanical and Manufacturing Engineering, Universiti Tun Hussein Onn Malaysia (UTHM) for their support.

Flexible flow shop scheduling problem with setup times and blocking constraint via genetic algorithm and simulation

Meng Zhang

Liaoning Equipment Manufacture College of Vocational Technology, Liaoning Shenyang ,China.
564202513@qq.com

Abstract. Flexible flow shop scheduling problem (FFSP) is recognized as an important class of problems in manufacturing systems. To consider more actual production factors, such as transportation time, setup times, blocking constraint, etc., we solved the FFSP using the improved genetic algorithm (GA) and discrete event simulation (DES). Firstly, a mathematical model for FFSP with the objective of minimizing total completion time was established. Besides, a GA mixed with Palmer heuristic algorithm was proposed to solve the mathematical model. Moreover, a DES software-plant simulation-was used to establish a more realistic production model, and an actual production workshop was modeled and optimized. After optimization, the total completion time and equipment utilization were greatly improved compared with the original scheduling. It shows the effectiveness and rationality of the proposed method to solve the actual FFSP.
Keywords: flexible flow shop scheduling problem (FFSP); genetic algorithm (GA); Plant Simulation; blocking; setup times.

1. Introduction

The flexible flow shop scheduling problem (FFSP) is the promotion of the flow shop scheduling problem (FSP). The FFSP is considered to have a few parallel production machines or lines. Flexible flow shop methodologies have been used in different industries such as semiconductor, automotive, chemical, etc. [1]

The FFSP has attracted considerable attention since it was identified in the 1970s [2] and many algorithms are proposed to solve the FFSP. Zhang et al. [3] proposed an ant colony optimization for permutation flow shop scheduling problem. Han et al. [4] studied the NSGA-II algorithm for multi-objective scheduling problem in a hybrid flow shop. And iterated greedy algorithms (IG) was used for the HFSP with total flow time criterion [5]. Moreover, the variable neighborhood search algorithm and meta-heuristics [6] was also used to solve the FFSP.

In addition to these algorithms, the genetic algorithms (GA) have been widely used in solving the FFSP for its powerful global search capabilities. Mousavi et al. [7] solved a bi-objective hybrid flow shop scheduling problem with the objective of minimizing the maximum completion time (makespan) and total tardiness using GA. Dabiri et al. [8] studied the multi-machine flow shop scheduling problems with rejection using GA. Wu et al. [9] proposed a hybrid non-dominated sorting genetic algorithm with variable local search to solve a multi-objective flexible flow shop scheduling problem which considers variable processing time. Luo et al. [10] proposed a dual heterogeneous island genetic algorithm for solving large size flexible flow shop scheduling problems. Azami [11] proposed a GA with a new cross-over operator and a heuristic algorithm to solve the FFSP, and considered the buffer capacity and the limited waiting time between the stations.

Discrete event simulation (DES) is a discrete-state and event-driven method for systems whose behavior is characterized by abrupt changes in the value of its state [12]. DES can efficiently build the simulation models of production systems considering more practical production factors. The Plant simulation software package is one of the DES software. Plant simulation has powerful logistics modeling and simulation capabilities to fully simulate the actual production process and can achieve precise and flexible control for different production conditions[13].

Based on the above literature, it can be found that various algorithms have been proposed to solve the FFSP. However, when using these intelligent algorithms, many actual production factors, such as blocking constraint and setup times are often neglected. Although a few researchers have considered some realistic factors[14], the establishment and solution of mathematical models are complicated. Moreover, although DES can efficiently establish a simulation model for production processes, few researchers have used the intelligent algorithms and DES in combination to solve the FFSP.

In this paper, the accurate simulation model of FFSP is established by Plant Simulation software and optimized by the improved GA. The actual production factors, such as the transportation time, the blocking constraint, and the setup times were considered in the simulation model. In addition, Palmer heuristic algorithm was used to generate an initial population, thereby speeding up the GA convergence. Finally, an actual flexible flow shop was used as a case to demonstrate the effectiveness of the proposed method in solving the FFPS.

2. Problem description and mathematical model

FFSP can be described as follows: n parts need to be processed, each part have m steps, and at least one of these processes contains parallel stations. The purpose of the study was to determine the order in which each part was machined at each station to minimize the total completion time of the batch.

The objective function is formulated as:

$$\min Z = C_{max} \tag{1}$$

Subject to:

$$\sum_{k \in K_j} x_{i,j,k} = 1, \quad \forall i, j \tag{2}$$

$$\sum_{t=1}^{n} y_{i,k,t} = x_{i,j,k}, \quad \forall i, j, k \in K_j \tag{3}$$

$$\sum_{i=1}^{n} y_{i,k,t} \le 1, \quad \forall k, t \tag{4}$$

$$\sum_{i=1}^{n} y_{i,k,t} \ge \sum_{i'=1}^{n} y_{i',k,t+1}, \quad \forall k, t < n \tag{5}$$

$$F_{i,j} = B_{i,j} + \sum_{k \in K_j} (P_{i,k} \times x_{i,j,k}), \quad \forall i, j \tag{6}$$

$$F_{k,t} = S_{k,t} + \sum_{i \in I} \sum_{j \in J} (P_{i,k} \times y_{i,k,t}), \quad \forall k, t \tag{7}$$

$$S_{k,t} \le B_{i,j} + M\_V \times \left(1 - y_{i,k,t}\right), \quad \forall i, j, k \in K_j, t \tag{8}$$

$$S_{k,t} + M\_V \times \left(1 - y_{i,k,t}\right) \ge B_{i,j}, \quad \forall i, j, k \in K_j, t \tag{9}$$

$$F_{k,t} \le S_{k,t+1}, \quad \forall k, t < n \tag{10}$$

$$E_{i,j} \le B_{i,j+1}, \quad \forall i, j < S \tag{11}$$

$$C_{\max} \ge E_{i,S}, \quad \forall i \tag{12}$$

$$x_{i,j,k}=\begin{cases}1, \text{ if } j\text{th job of part } i \text{ is processed in machine } k\\ 0, \text{ otherwise}\end{cases} \quad (13)$$

$$y_{i,k,t}=\begin{cases}1, \text{ if part } i \text{ is processed in the } t\text{th job of machine } k\\ 0, \text{ otherwise}\end{cases} \quad (14)$$

Where i denotes part, I denotes a set of parts, J denotes a set of jobs, K denotes a set of machines, P_{ik} denotes the processing time of part I in the j th event, S_{kt} denotes the start time of t th event in machine k ; F_{kt} denotes the end time of t th event in machine k ; $C_{\max}$ denotes the maximum completion time; $B_{i,j}$ denotes the start time of the jth event of part i; $E_{i,j}$ denotes the end time of the jth event of part i; V is a sufficiently large positive number.

Eq. 2 ensures that any process of the part can only be processed once; Eq. 3 ensures that whether a part is processed on this machine is determined by the assignment scheme; Eq. 4 ensures that a machine can only process one part at a time; Eq. 5 ensures that each machine is processed in the specified order; Eq. 6 and Eq. 7 ensure that the end time of each operation equals the start time plus the processing time; Eq. 8 and Eq. 9 ensure that the start time of the machine is equal to the start time of the part processed in the machine; Eq. 10 ensures that the end time of the previous event in any machine is not greater than the start time of subsequent events; Eq. 11 ensures that the end time of the previous event of a part is not greater than the start time of the subsequent event; Eq. 12 represents the Makespan.

3. Design of Genetic Algorithms

GA is a highly parallel, stochastic and adaptive optimization algorithm. It represents a solution and an objective function of the problem using chromosomes and the fitness of these chromosomes. The basic elements of GA include chromosome coding, population initialization, decoding, selection, crossover and mutation. Firstly, a group of initial population representing the solution of the problem is generated by coding, and the chromosomes with good adaptability are selected according to its fit-ness value. These chromosomes are regarded as the paternal generation. Then, through the crossover and mutation, excellent genes are inherited to form offspring population. Through the continuous evolution of the chromosome, the population eventually converges to the optimal individual and gets the optimal solution of the problem.

3.1. *Coding and Decoding*

The coding method was based on process expression, the assembly sequence of each product is encoded as a chromosome. Each gene represents a process, and for the same part, the sequence in the chromosome represents the assembly sequence of the parts. After decoding, O_{ijk} indicates that the j process of product i operates on the workstation k .This coding and decoding method is illustrated with a scheduling problem of three production workstations with three products. The operation schedule of the product is shown in Table 1. Each product is processed on three workstations in turn: SP1, SP2 and SP3.If a scheduling code is $[1\ 2\ 1\ 1\ 3\ 2\ 3\ 2\ 3]$, it can be decoded as [$O_{111}O_{211}O_{122}O_{133}O_{311}O_{222}O_{322}O_{233}O_{333}$], according to the encoding and decoding rules. The decoding process and scheduling schematics are shown in Figure 1 and Figure 2, respectively.

Table 1. Processing time of three production workstations with three products.

Product	Processing Time		
	SP1	SP2	SP3
J1	2	2	4

J2	3	4	3
J3	3	4	5

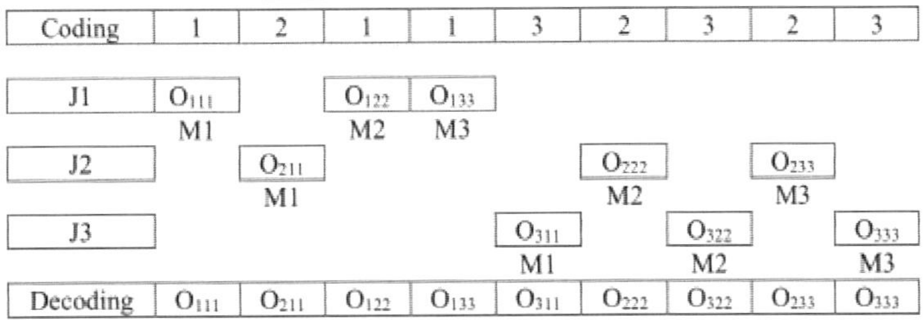

Figure 1. The decoding process based on process expression.

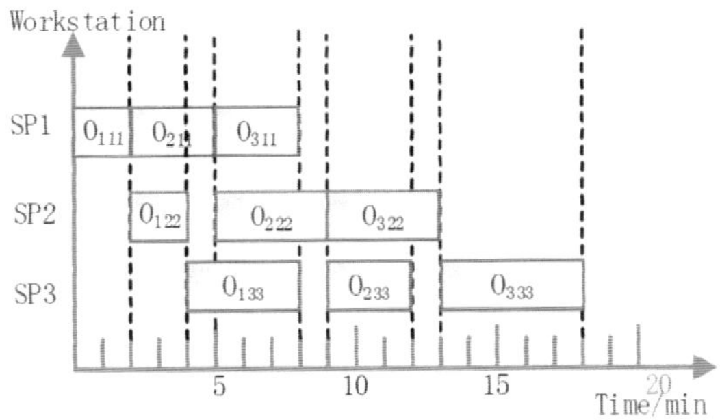

Figure 2. Scheduling Diagram.

3.2. Initial population.
The generation of an initial population is usually random. In this paper, a Palmer heuristic algorithm was used to produce an initial population. The Palmer algorithm ranks the jobs in the order that the slope index does not increase. The slopes of the parts are as follows:

$$\lambda_i = \sum_{j=1}^{m} [\frac{j-(m+1)}{2}] t_{ij} \quad (i = 1,2,...,n) \tag{15}$$

Where M denotes the number of machines; t_{ij} denotes the processing time of part J_i on machine M_j.

3.3. Selection.
Selection operations select a number of better individuals from the current population with a certain probability. Individuals with high fitness are more likely to be selected. Selection operation makes the high-performance individuals have a greater survival probability, avoids the deletion of effective genes, and improves the global convergence and computational efficiency of the solution.

3.4. Crossover.
Crossover operation selects two individuals from the parent population for gene exchange to form a new generation of individuals. Partial mapping cross (PMX) approach was used in this paper. Firstly, two random points are randomly inserted into the two parents to exchange the fragments between the intersections of the parents. Then, for the new individuals formed after the exchange, inspect whether the gene exchanged conflicts with the genes outside the chromosome crossing point. And the conflicting genes will be modified by partial mapping.

4. Simulation Modeling and Optimization
An actual production workshop was taken as an example, and the plant simulation software was used to establish the simulation model of the workshop production. The scheduling optimization model of the

assembly workshop was established, according to the assembly process and plant layout. As shown in Figure 3, the assembly shop includes a material shop, two assembly shop, and a packaging workshop. The AS workstation has two parallel workstation s, and the T workstation has three parallel workstations. The assembly assignment on the parallel machine is assigned according to the machine's no idle principle. The Automated Guided Vehicle (AGV) is used for transportation between the workshops, and the transmission lines are used between workstations. Five types of products are produced, and the operation time for each product are shown in Table 2.

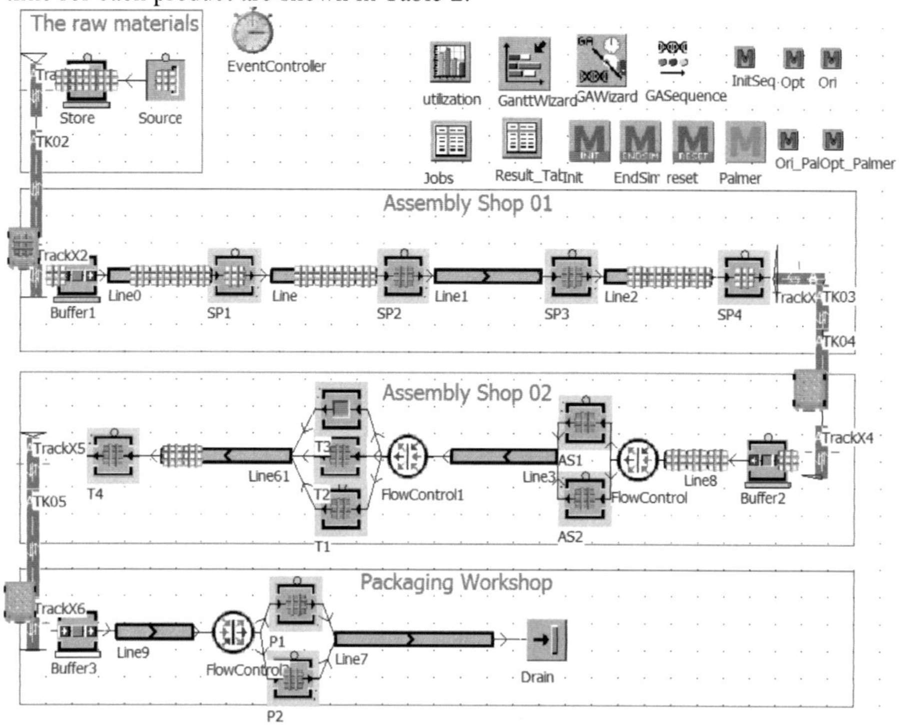

Figure 3. Production simulation model of an assembly shop.

Table 2. The operation time (s) for each product.

Product	Workstation											
	SP1	SP2	SP3	SP4	AS1	AS2	T1	T2	T3	T4	P1	P2
J1	63	75	81	76	297	246	400	328	247	122	215	315
J2	88	90	74	122	143	372	303	347	267	143	285	250
J3	152	130	105	116	190	262	352	273	413	230	175	470
J4	110	133	125	132	375	327	250	373	237	185	195	310
J5	133	115	113	141	287	205	377	283	340	113	263	390

There are setup times when the AS workstation processes different parts. The required setup times are shown in Table 3 and Table 4.

Table 3. Setup times (s) for different products in AS1 workstation.

Product	Product				
	J1	J2	J3	J4	J5
idle	18	13	19	17	14
J1	0	11	21	5	6
J2	12	0	17	5	5
J3	5	5	0	10	9
J4	4	5	4	0	4
J5	5	8	5	5	0

Table 4. Setup times (s) for different products in AS2 workstation.

Product	Product				
	J1	J2	J3	J4	J5
idle	13	16	15	22	19
J1	0	15	16	5	9
J2	6	0	5	4	7
J3	7	4	0	9	6
J4	8	7	4	0	8
J5	8	8	5	6	0

5. Establishment of the simulation model

An actual production workshop was taken as an example, and the plant simulation software was used to establish the simulation model of the works.Assuming that the quantity of each product is one, the simulation is run according to the initial production sequence (J_1-J_5). After the simulation, the total assembly completion time is 40 minutes and 4 seconds, and the Gantt chart of each workstation is shown in Figure 4.

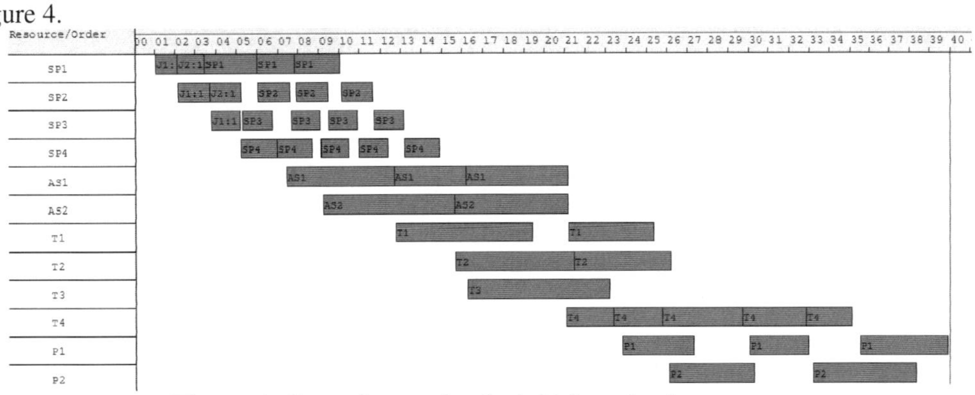

Figure 4. Gantt chart under the initial production sequence.

The scheduling simulation model was optimized by the improved GA. The number of generations is 20, the generation size is 30, and the mutation probability is 0.1.

After optimization, the total completion time is 36 minutes and 15 seconds, and the optimal assembly scheduling sequence is J_2-J_1-J_4-J_5-J_3. The evolution curve and the fitness of the offspring are shown in Figure 5 and Figure 6, respectively. It can be seen from the figure that as the evolutionary algebra increases, the result tends to be optimal and converges to the optimal scheduling sequence after the seventh generation. The Gantt chart under optimal production sequence is shown in Figure 7.

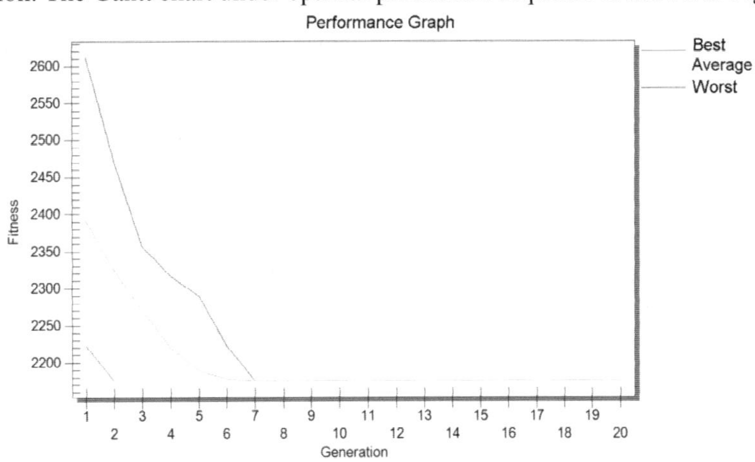

Figure 5. Evolution curve of the GA.

Figure 6. Fitness of the offspring.

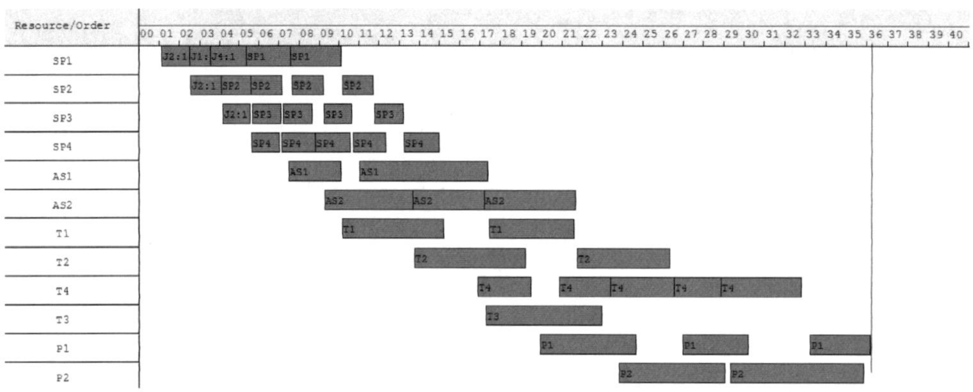

Figure 7. Gantt chart under the optimal production sequence.

In addition, under stable production conditions, the facility utilization corresponding to the initial production sequence and the optimized production sequence are shown in Figure 8 and Figure 9, respectively. It can be seen from the figure that after optimization the utilization of each workstation is increased.

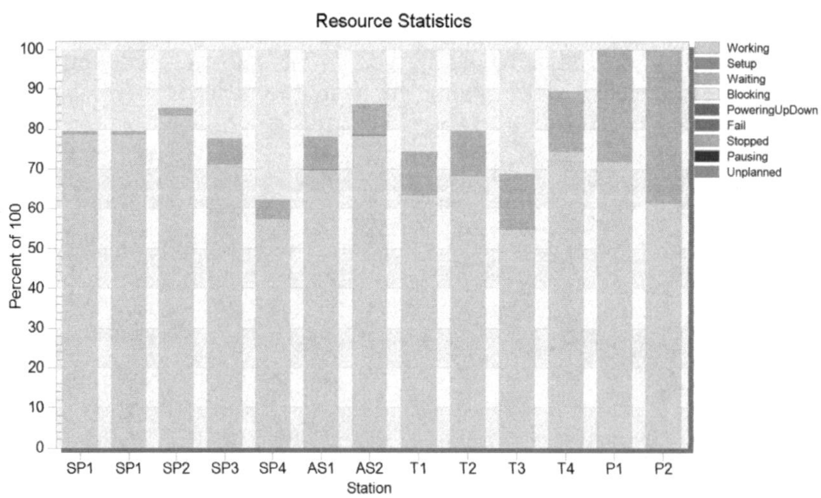

Figure 8. Facility utilization under the initial production sequence.

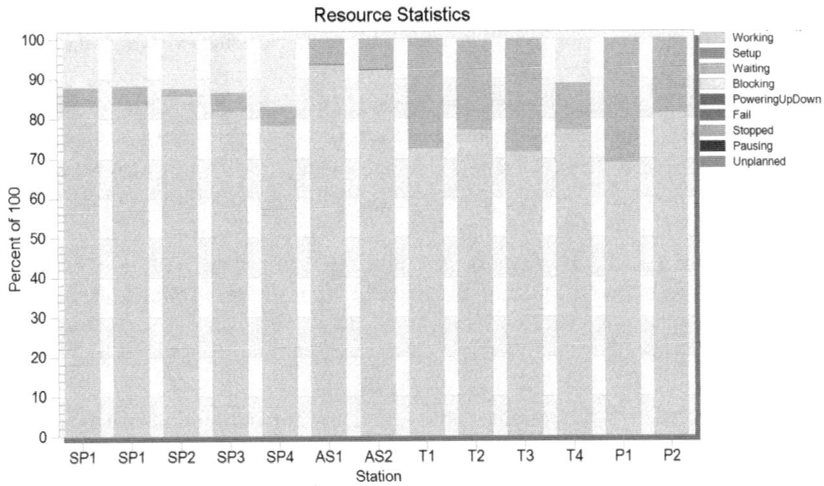

Figure 9. Facility utilization under the optimal production sequence.

6. Conclusions

To obtain a more realistic simulation model and optimization results for FFSP, this paper uses a combination of improved genetic algorithm and discrete event simulation. The Palmer heuristic algorithm was used to obtain the initial solution, the GA was used to obtain the optimal solution, and the Plant Simulation software was used for production simulation. The process based coding method was adopted to prevent the generation of illegal chromosomes, and encoding and decoding are simpler and more intuitive. In addition, more realistic production factors, such as transportation time, setup times and blocking constraints are considered in the Plant Simulation software. After optimization, total completion time is reduced and equipment utilization is increased. It shows that the optimization method proposed in this paper can effectively solve the FFSP with more realistic constraints, and can provide decision guidance for the scheduling optimization in actual production conditions.

References

[1] LEE T-S, LOONG Y-T. A review of scheduling problem and resolution methods in flexible flow shop [J]. International Journal of Industrial Engineering Computations, 2019, 67-88.

[2] CHOI S H, WANG K. Flexible flow shop scheduling with stochastic processing times: A decomposition-based approach [J]. Computers & Industrial Engineering, 2012, 63(2): 362-73.

[3] ZHANG Y, YU Y, ZHANG S, et al. Ant colony optimization for Cuckoo Search algorithm for permutation flow shop scheduling problem [J]. Systems Science & Control Engineering, 2018, 7(1): 20-7.

[4] HAN Z, WANG S, DONG X, et al. Improved NSGA-II algorithm for multi-objective scheduling problem in hybrid flow shop [M]. Innovative Techniques and Applications of Modelling, Identification and Control. Springer. 2018: 273-89.

[5] ÖZTOP H, TASGETIREN M F, ELIIYI D T, et al. Iterated greedy algorithms for the hybrid flowshop scheduling with total flow time minimization [J]. 2018, 379-85.

[6] ABDOLLAHPOUR S, REZAIAN J. Two new meta-heuristics for no-wait flexible flow shop scheduling problem with capacitated machines, mixed make-to-order and make-to-stock policy [J]. Soft Computing, 2016, 21(12): 3147-65.

[7] MOUSAVI S M, MAHDAVI I, REZAEIAN J, et al. An efficient bi-objective algorithm to solve re-entrant hybrid flow shop scheduling with learning effect and setup times [J]. Operational Research, 2016, 18(1): 123-58.

IOP Conf. Series: Materials Science and Engineering **637** (2019) 012007 doi:10.1088/1757-899X/637/1/012007

[8] DABIRI M, DARESTANI S A, NADERI B. Multi-machine flow shop scheduling problems with rejection using genetic algorithm [J]. International Journal of Services and Operations Management, 2019, 32(2): 158-72.
[9] WU X, SHEN X, CUI Q. Multi-Objective Flexible Flow Shop Scheduling Problem Considering Variable Processing Time due to Renewable Energy [J]. Sustainability, 2018, 10(3): 841.
[10] LUO J, EL BAZ D. A Dual Heterogeneous Island Genetic Algorithm for Solving Large Size Flexible Flow Shop Scheduling Problems on Hybrid Multicore CPU and GPU Platforms [J]. Mathematical Problems in Engineering, 2019, 1-13.
[11] AZAMI A, DEMIRLI K, BHUIYAN N. Scheduling in aerospace composite manufacturing systems: a two-stage hybrid flow shop problem [J]. The International Journal of Advanced Manufacturing Technology, 2018, 95(9-12): 3259-74.
[12] SILVA M. On the history of Discrete Event Systems [J]. Annual Reviews in Control, 2018, 213-22.
[13] FERRO R, CORDEIRO G A, ORDOñEZ R E C. Dynamic Modeling of Discrete Event Simulation [J]. 2018, 248-52.
[14] JANARDHANAN M N, LI Z, BOCEWICZ G, et al. Metaheuristic algorithms for balancing robotic assembly lines with sequence-dependent robot setup times [J]. Applied Mathematical Modelling, 2019, 256-70.

IOP Conf. Series: Materials Science and Engineering **637** (2019) 012008 doi:10.1088/1757-899X/637/1/012008

Optimization Analysis of Minimum Flat Curve Radius of High Speed Maglev Line with Design Speed of 500 km/h

Fujie Jiang[1], Xiangdong Sun[1] Huibai Li[2], Jingyu Huang[2], Dongshuai Li[2]

[1]CRRC QINGDAO SIFANG CO.,LTD, Qingdao 266000,China.
[2]School of Tongji University, Shanghai 201804,China.
Email: huibai_li@163.com

Abstract. Firstly, based on the traditional circuit design method, the theoretical calculation value of the minimum radius of the flat curve under the design speed of 500 km/h is calculated in this paper; Through the analysis of the structural characteristics of the TR08 high-speed maglev train and the degree of freedom of each component, a 129-DOF high-speed maglev vehicle-line dynamics model is established by using the multi-body dynamics software SIMPACK; And using the modle to simulate and analyze the influence of the radius of the flat curve on the dynamic performance of the vehicle. Finally, according to the simulation results, the theoretical calculation of the minimum radius of the flat curve is optimized and adjusted, and the minimum radius recommended value of the flat curve that satisfies the dynamic evaluation index is obtained. After the simulation optimization, the unbalanced lateral acceleration and the vertical acceleration dynamic response of the vehicle passing through the designed curve section are significantly reduced, and the ride comfort of the passengers is good.

1. Introduction

The traditional wheel-rail trains have reached the speed bottleneck of 400 km/h due to wheel-rail contact restrictions. If the speed is further increased, it will bring out a huge increase in operating costs. The high-speed maglev system has got rid of the wheel-rail contact restriction and obtained further speed-up space. For example, the German TR maglev train has achieved a test speed of 505 km/h, so the high-speed maglev system is considered to be the development direction of the next generation of rail transit The high-speed maglev transportation system includes vehicles, lines, traction power supply, operation control and other major parts [1], among which the line part is one of the most important parts of the high-speed maglev transportation system. The quality of line design is directly related to the operational safety, stability and ride comfort of the maglev train [2]. Traditionally, the line selection design is mostly based on the static method. The vehicle is regarded as a particle or a rigid body [3], and the theoretical calculation value of the minimum horizontal curve radius of the line is obtained on the basis of the unbalanced lateral acceleration limit value of the vehicle body. The method does not take into account the complex structure of the vehicle and the relative motion between the components is considered less. The calculated minimum radius of the flat curve may not meet the dynamic performance requirements of the high-speed maglev vehicle. Therefore, in order to ensure the safety of train operation and passenger ride comfort, it is particularly important to verify and optimize the value of the minimum radius parameters of the flat curve under the design speed of 500 km/h calculated by traditional circuit design method from the dynamic point of view. At the same time, it also has certain innovative significance in the design of high-speed maglev lines.
Theoretical Calculation of the Minimum Radius of the Flat Curve of High-Speed Maglev Line

Simulation Model. The minimum radius of the flat curve is one of the main technical standards for line selection design. In addition to the comfort requirement, the minimum radius of flat curve of the traditional wheel-rail trains should also meet the safety requirement to prevent derailment and overturning accidents due to mismatched line design parameters. For the high-speed maglev system, the vehicle adopts the outsourced suspension frame structure, and the vehicle always surrounds the track, which greatly reduces the possibility of derailment and overturning of the vehicle [4]. Therefore, based on the structure characteristics and engineering technical requirements of high-speed maglev train, the main factors affecting the minimum radius of flat curve of high-speed maglev line are the vehicle structure requirements and passenger ride comfort requirements [5].

（1）Vehicle construction requirements

The structural characteristics of TR08 maglev train determine that the minimum curve radius of the train passing through the flat curve cannot be lower than 350m.

（2）Passenger ride comfort requirements

The force diagram of the maglev train running on the curve is shown in Fig.1.

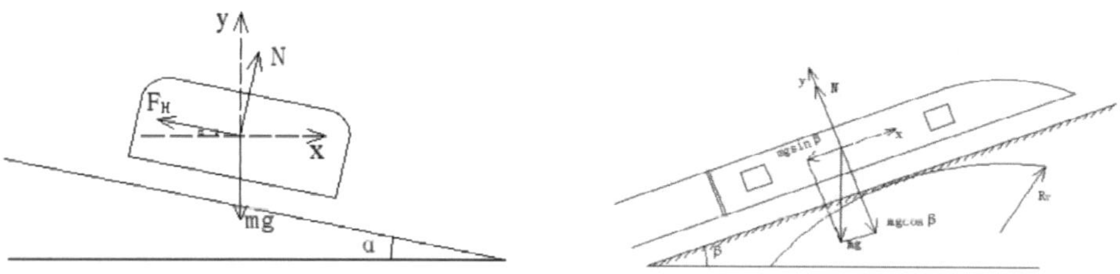

(a) flat curve (b) vertical curve

Figure 1. Force analysis of high-speed maglev train running in curved section

When the train runs in the curve section, if the designed driving speed does not match the setting value of the cross-slope angle of the line, the train will generate a certain amount of centrifugal force, and the passengers in the train will bear the unbalanced lateral acceleration [6]. According to the passenger's ride comfort requirements, this unbalanced lateral acceleration cannot exceed a certain limit, that is:

$$a_y \le a_{max}. \tag{1}$$

In this equation: a_y ——unbalanced lateral acceleration (m/s²)

a_{max} ——limit value of unbalanced lateral acceleration (m/s²)

According to the force diagram of the train in Fig. 1, the unbalanced lateral acceleration in the horizontal direction can be expressed as follows:

$$a_y = \frac{F_H \cos\alpha - N \sin\alpha}{m}. \tag{2}$$

The centrifugal force F_H and suspension force N that the train receives during the running of the curved section can be expressed as follows:

$$F_H = m\frac{\left(\frac{V}{3.6}\cos\beta\right)^2}{R_H}. \tag{3}$$

$$\mathrm{N} = \mathrm{mg}\cos\beta - m\frac{(V/3.6)^2}{R_V}. \tag{4}$$

The reason why the number 3.6 appears in Eq.3 is because the speed (km/h) needs to be converted to the international unit speed (m/s). Bringing the Eq.3 and Eq.4 into the Eq.2 and transforming the Eq.2, the equation for calculating the radius of the flat curve can be expressed as follows:

$$R_H = \left|\frac{(V/3.6)^2 \cos\alpha \cos^2\beta}{a_y + \left[g\cos\beta + m\frac{(V/3.6)^2}{-R_V}\right]\sin\alpha}\right|, a_y \le a_{max}. \tag{5}$$

In this equation: R_H——Flat curve radius(m);

R_V——Vertical curve radius（m）；

V——Train running speed（km/h）；

α ——the Angle of the cross slope（°）；

β ——Longitudinal gradient（‰）；

g——Gravity acceleration，9.81 m/s^2;

In practical engineering, the order of circuit design is generally to firstly design a flat curve and then design a vertical curve. In the design of flat curve, the influence of longitudinal slope and the radius of vertical curve are generally not considered. Therefore, taking the longitudinal slope $\beta = 0$ and the vertical curve radius R_V=∞, the Eq.5 can be simplified to the following equation:

$$R_H = \left|\frac{(V/3.6)^2 \cos\alpha}{\alpha_y + g \times \sin\alpha}\right|, a_y \leq a_{max}. \tag{6}$$

Regarding to the value of the cross slope angle α, referring to the relevant provisions of the Shanghai High Speed Maglev Demonstration Operation Line and the German Transrapid Line Selection Instruction: the maximum cross-slope angle of high-speed maglev line shall not exceed 12 degrees, and in extreme cases, the maximum 16 degrees cross-slope angle can be obtained after the technical and economic demonstration.

Therefore, with the values of each parameters substituted into the above equation, the calculated minimum radius of the flat curve of the high-speed maglev line that satisfies the ride comfort requirement under the design speed of 500 km/h can be calculated, as shown in Table 1.

Table 1. Calculated minimum radius of flat curve of high-speed maglev line at the speed of 500 km/h

Design Speed（km/h）	Minimum Radius of Flat Curve	
	General situation	Difficult situation
500	6400	5800

Note：1.The general situation in the table refers to the unbalanced lateral acceleration limit of 1.0 m/s^2 and 1.25 m/s^2 in difficult cases.

2.When calculating, the maximum cross slope angle is taken as 12 degrees.

Establishment of High Speed Maglev Vehicle-Line Dynamic Model

In order to verify and optimize the theoretical calculation value of the minimum radius of the flat curve under the designed driving speed of 500 km/h, a complete vehicle-line dynamic model must be established. By setting different simulation conditions, the law of influence of the change of the radius of the flat curve on the dynamic performance of the vehicle can be simulated and analyzed.

The research is based on the TR08 maglev train. One TR08 high-speed maglev vehicle contains one body, four levitation frame running mechanisms, 14 suspension electromagnets, 12 guide electromagnets, 2 brake electromagnets, 16 anti-roll bolsters, 16 pendulum and other major components [7]. When considering the influence of the change of the radius curve parameter on the dynamic performance of the vehicle, only the uniform operation conditions is considered, and the traction and braking conditions are not involved, and the longitudinal motion of the vehicle is not considered. Therefore, the guide electromagnet can be used to replace the position of the brake electromagnet and play the guiding function in the process of establishing the model. By analyzing the structural characteristics and connection relationships of each component of the vehicle, the degree of freedom of each component of a single vehicle can be obtained as shown in Table 2.

Table 2. Freedoms of Components of a Single Vehicle

Components	Transverse	Sinking	Rolling	Nodding	Shaking head
Car body	√	√	√	√	√
Suspension electromagnet	-	√	-	√	-
Guided electromagnet	√	-	-	-	√
Bolster	-	-	√	-	-
Pendulum	-	-	√	√	-

According to the degree of freedom of each component and based on the principle of multi-body dynamics, the models of each component can be established in SIMPACK software platform. The connection and force transmission relationships between components can be replaced by hinges, force elements and constraints, and the line model can be obtained by setting the line parameters in Track module [8]. Finally, a high-speed maglev vehicle-line dynamic model with 129 degrees of freedom is obtained through a series of steps, as shown in Fig. 2.

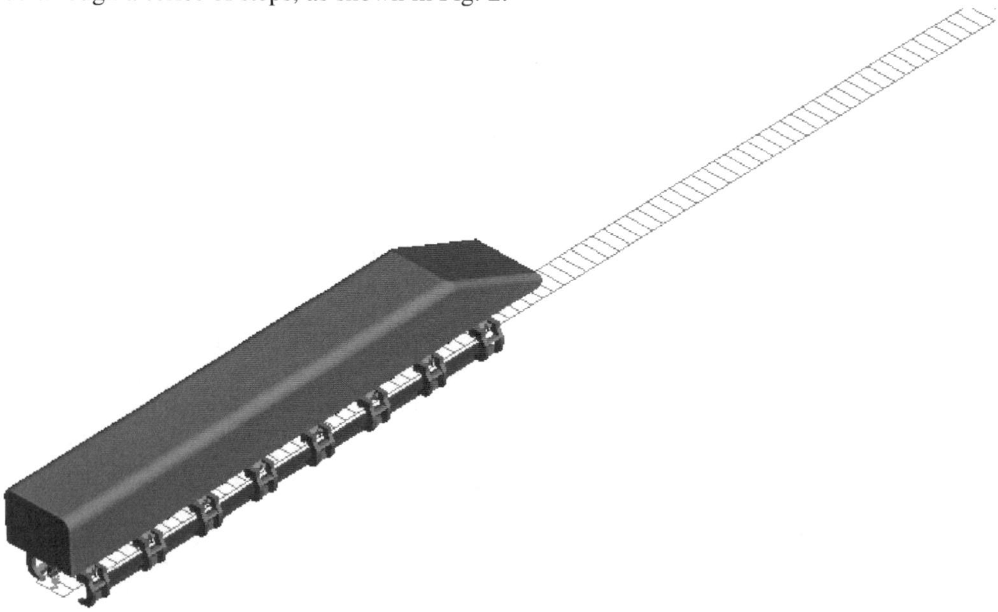

Figure 2. High Speed Maglev Vehicle-Line Dynamic Model

2. Model Simulation Analysis

Evaluation Index. When the maglev train runs on the curve, it will generate a certain amount of acceleration and vibration. The passengers on the train have a certain tolerance limit for the perceived acceleration and vibration. This limit must meet the requirements of dynamic performance. Up to now, the high-speed maglev transportation system has not yet formed a set of evaluation indicators to evaluate the dynamic performance of vehicles. Referring to the rail transit system, the ISO2631 standard is usually used to evaluate the relationship between passenger comfort and vehicle acceleration.

The ISO2631 standard is used to evaluate the human body's ability to withstand vibration. After extensive testing and demonstrations, it has become an international general evaluation standard applicable to various industries. In this study, the ISO2631 standard is used as an evaluation index for evaluating the dynamic performance of Maglev vehicles. Its standard is shown in Table 3

Table 3. ISO2631 Comfort Evaluation Index [9]

Acceleration Transmissibility (m/s^2)	Comfort Condition	Comfort Level
<0.315	Very comfortable	Level 1
0.315～0.63	Comfortable	Level 2
0.5～1.0	Fairly comfortable	Level 3
0.8～1.6	Discomfort	Level 4
1.25～2.5	Very uncomfortable	Level 5
>2.5	Extremely uncomfortable	Level 6

When simulating the dynamic response of vehicle with the change of the radius of flat curve, the corresponding comfort level can be determined by comparing the unbalanced lateral acceleration and the vertical acceleration in the simulation results with the comfort evaluation standard in Table 3. The design requirements for high-speed maglev line selection are as follows: in the case of general curve sections, the passenger's ride comfort level is not less than level 2, that is, "comfortable"; in the case of difficult curve sections, it is not less than level 3, that is, "fairly comfortable".

Simulation and Analysis of Vehicle Dynamic Performance with the Change of Radius Parameters of Flat Curve at the Speed of 500 km/h. Different working conditions are set up in SIMPACK simulation platform to simulate the influence of different flat curve radius on vehicle dynamic performance at 500 km/h speed. In the simulation, the unbalanced lateral acceleration and vertical acceleration dynamic response simulation results of vehicle body center are taken as the evaluation parameters to evaluate vehicle dynamic performance. After a series of simulation, the simulation results are shown in Table 4. The relationship between the lateral acceleration, the vertical acceleration and the radius of the flat curve is shown in Fig. 3

Table 4. simulation results of vehicle-line dynamic model under different radius of flat curve at the speed of 500 km/h

Radius of flat curve（m）	Unbalanced lateral acceleration （m/s²）	Vertical acceleration （m/s²）
5800	1.185	0.707
6000	1.077	0.683
6200	0.975	0.662
6400	0.881	0.641
6600	0.791	0.621
6800	0.708	0.602
7000	0.627	0.586
7200	0.552	0.570

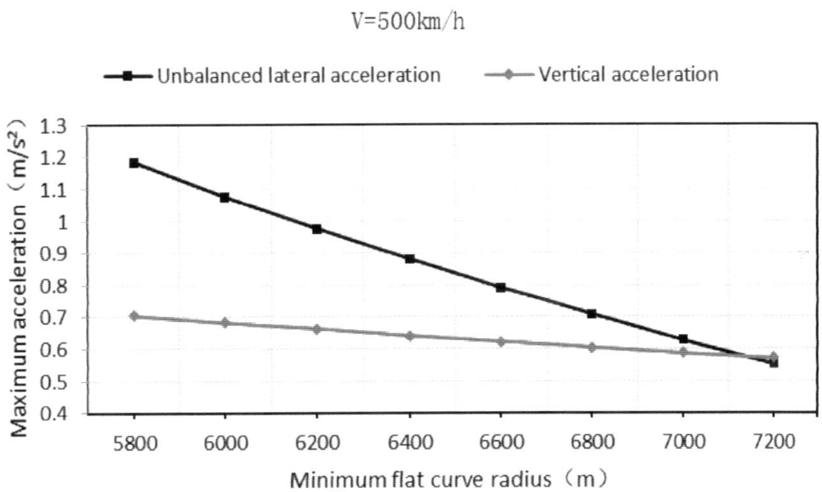

Figure 3. Relationship between the lateral, vertical acceleration and the flat curve radius

Observing the curve trend of Fig.3, we can see that the unbalanced lateral acceleration and vertical acceleration of the vehicle during the running of the curve section are all related to the radius of the flat curve, and both decrease with the increase of the radius of the curve. Careful observation of the trend of the curve slope shows that the radius of the slope of the shows that the radius of the flat curve has a greater impact on the lateral acceleration. The lateral acceleration first decreases rapidly with the increase of the radius of the flat curve. When the radius of the curve increases to a certain extent, the lateral acceleration decreases more slowly with the increase of the radius of the flat curve.

Combined with the data in Table 4, when the theoretical calculated radius of the flat curve is 5800m, the lateral acceleration is 1.185 m/s², and the vertical acceleration is 0.707 m/s² and the lateral comfort is level 4, the "uncomfortable" state, and the vertical comfort is level 3, that is "fairly comfortable"state, at this time, the lateral comfort does not meet the comfort requirements. When the radius of flat curve is 6400m, the lateral acceleration is 0.881 m/s², and the lateral comfort is level 4, which still does not meet the requirements of lateral comfort. When the radius of the flat curve increases to 6600m, the lateral acceleration is 0.791 m/s² and the lateral comfort is level 3, which is the "fairly comfortable" state, and the vertical acceleration is 0.621 m/s² and the vertical comfort is level 2, which is the "comfortable" state, so the lateral and vertical comfort meet the comfort requirements under difficult circumstances, and can be used as the minimum radius of the flat curve under difficult situation. When the radius of flat curve is further increased to 7000m, the value of lateral acceleration is 0.627 m/s² and the lateral comfort is level 2, which is "comfortable" state. And the value of vertical acceleration is 0.586 m/s² and the vertical comfort is level 2, the "comfortable" state. So the horizontal and vertical comfort meet the comfort requirement under general situations, and can be taken as the minimum radius of flat curve under general situations. Therefore, from the perspective of comfort requirement, the minimum radius of flat curve that satisfies passengers' comfort requirement should be no less than 7000m under general situations and not less than 6600m under difficult situations.

In summary, the minimum radius of the flat curve of 6400m under general situations and 5800m under difficult situations calculated by traditional circuit design method cannot meet the passenger comfort requirements. After dynamic simulation optimization, its value should be adjusted to 7000m under general situations and 6600m under difficult situations, as shown in Table 5.

Table 5. Suggested values for the minimum radius of the flat curve of high-speed maglev line at the speed of 500 km/h

Designed Driving Speed (km/h)	Minimum Radius of Flat Curve（m）	
	General Situation	Difficult Situation
500	7000（6400）	6600（5800）

Note: The theoretical calculation value is in brackets, and the simulation optimization value is outside brackets.

After the optimization, when the high-speed maglev train with the maximum design speed of 500 km/h passes through the flat curve section under normal and difficult situations, the lateral and vertical comfort of the car body can meet the requirements, and the passengers' ride comfort is good. And through simulation and optimization, the lateral acceleration and vertical acceleration dynamic response of the car body passing through the curve section are reduced by 33.2% and 12.2% under general situations, and 28.8% and 8.6% under difficult situations, which better improves the ride comfort of passengers.

3. Conclusion

In this paper, by establishing the high-speed maglev vehicle-line dynamic model, the theoretical calculation value of the minimum radius of flat curve under the condition of speed 500 km/h obtained by the traditional circuit design method is simulated and optimized, and the ride comfort of passengers is taken as the evaluation index to evaluate the dynamic performance of maglev vehicle under different flat curve radius conditions. Finally, according to the simulation results, the theoretical calculation value of the minimum radius of the flat curve is optimized and adjusted, and the recommended value of the minimum radius of the flat curve of the high-speed maglev line with a design speed of 500 km/h that meeting the ride comfort of passengers is obtained. The optimization results show that:

（1）After the optimization, when the high-speed maglev train with the maximum design speed of 500 km/h passes through the flat curve section under normal and difficult situations, the lateral and vertical comfort of the car body can meet the requirements, and the passengers' ride comfort is good.

（2）Through simulation and optimization, the lateral acceleration and vertical acceleration dynamic response of the car body passing through the curve section are reduced by 33.2% and 12.2% under general situations, and 28.8% and 8.6% under difficult situations.

Acknowledgments

This work was financially supported by National "13th Five-Year" Plan for Science & Technology Support 'Research on High-Speed Maglev Transportation Engineering key technology' (2016YFB1200602) fund.

IOP Conf. Series: Materials Science and Engineering **637** (2019) 012008 doi:10.1088/1757-899X/637/1/012008

References

[1] Wu Xiangming. Maglev train [M]. Shanghai: Shanghai Science and Technology Press, 2003.

[2] Li Huibai, Huang Jingyu. Research on Maximum Longitudinal Slope Value of High Speed Maglev Line[J]. Urban Rail Transit Research, 2018, 21(11): 44-48.

[3] Zhou Hewang. Dynamics simulation analysis of high-speed railway circular curve parameters based on 3D line model [D]. Southwest Jiaotong University, 2014.

[4] Deng Xiaoxing. Research on Dynamics Performance of Medium and Low Speed Maglev Vehicle System[D]. Southwest Jiaotong University, 2009.

[5] Ministry of Housing and Urban-Rural Development of the People's Republic of China, National Development and Reform Commission of the People's Republic of China. High-speed Maglev Traffic Construction Standards (Trial) [S]. Jianbiao 161-2012.

[6] Shi Jin, Feng Yawei, School of Civil Engineering and Architecture, Beijing Jiaotong University, Beijing. Research on main technical parameters of flat vertical curve of high-speed maglev traffic line[J]. 2005 National Doctoral Academic Forum - Transportation Engineering, 2010.

[7] Xie Weimin. Research on simulation model of magnetic suspension vehicle-line coupling dynamics [D]. Southwest Jiaotong University, 2005.

[8] Li Xiaobin, Fang Yu, Jin Bo, et al. Research on Operational Safety Simulation and Stationarity of a Certain Urban Rail Transit Vehicle Based on SIMPACK Software[J]. Urban Rail Transit Research, 2015, 18(7): 85-89.

[9] Qiu Feijian. Research on high-speed railway line parameters based on passenger comfort [D]. Southwest Jiaotong University, 2008.

IOP Conf. Series: Materials Science and Engineering **637** (2019) 012009 doi:10.1088/1757-899X/637/1/012009

Stability and reinforcement analysis of weathering steel high voltage transmission tower

YAO Yao[1], WANG Ling Xu[2] and ZHANG You Jia[3]

[1]Power Dispatching Control Center of Guizhou Power Grid Co. Ltd., Guiyang 550002, China.
[2]Guizhou Power Grid Co. Ltd. Electric Science Research Institute, Guiyang 550001, China.
[3]School of Civil Engineering and Architecture, Northeast Electric Power University, Jilin 132012, China
E-mail: wlx43@qq.com

Abstract. The local instability of weathering steel transmission towers is an important reason for the collapse of transmission towers.In this paper, the wind speed is converted into wind pressure and applied to ANSYS finite element model established according to the actual equal proportion, and the influence of secondary stress and the orientation of Angle steel are considered.The stability of transmission tower is analyzed according to buckling coefficient.The results show that the buckling coefficient and the buckling position increase with the reinforcement degree of the main material of the weathering steel transmission tower.After the reinforcement of the first and second sections of the main transmission material is completed, the buckling position moves up to the web bar of the transverse arm.The buckling coefficient of T type is about 39% and 25% higher than that of the cross type and Z type respectively, which can effectively improve the safe operation ability of the line.

1. Introduction

Weathering steel transmission lines, as an important part of the power system, are widely distributed in various regions of China. They are affected by natural disasters such as freezing rain, snow and strong winds. The occurrence of severe faults such as local buckling or partial collapse of transmission lines seriously affects the safe and stable operation of power system[1].Due to the limitations of design codes at that time, transmission towers built in the early stage failed to fully consider the influence of extreme adverse natural environment, such as maximum design wind speed, variation coefficient of wind pressure height and wind carrier shape coefficient, etc., resulting in tower collapse accidents.In fact, strength failure and buckling are the main failure modes of transmission tower collapse[2,3].
In this paper, the reinforcement scheme that is connected by a single plate and combined with various cross-section forms is adopted. ANSYS software is used to model the actual tower, and the wind pressure is converted into the node load and the eigenvalue buckling analysis is carried out, so as to select the best reinforcement scheme.

2. Wind load calculation of transmission tower

The wind load mainly depends on the wind speed. The wind speed of the transmission tower should be converted to the wind pressure load, and the basic wind pressure load should be corrected by the working condition coefficient. According to the structural characteristics of the transmission tower, the wind pressure load of the cross arm and the wind pressure load of the tower should be calculated separately[4].

2.1. Wind speed and wind pressure

The Bernoulli equation in fluid mechanics can transform wind speed into reference wind pressure W_0

$$W_0=\frac{1}{2}\rho v^2=\frac{1}{2}\frac{\gamma}{g}v^2 \quad (1)$$

Where: represents the air particle density of the site environment (t/m^3) ; γ represents the gravitational density of a gas kN/m^3 ; v represents the wind speed (m/s) 。Combined with China's meteorological conditions, the design specification is usually taken $\gamma/(2g)=1600$ 。

2.2. Wind load standard value calculation

Transmission tower wind load is not only related to the reference wind pressure, but also takes into account the influence of tower height, component shape, fluctuating wind and icing[5]. When the wind direction is perpendicular to the structure surface, the formula for calculating the standard wind load is as follows:

$$W_s=\mu_z\mu_s\beta_z BA_f W_0 \quad (2)$$

3. Wind buckling analysis of transmission towers

When the loading test is carried out on the structure, large deformation will occur suddenly when the critical value is reached. At this time, even if the members continue to apply negligible small load, the existing equilibrium state of the structure is immediately broken, which is accompanied by a large deformation phenomenon known as structural instability or buckling[6].

3.1. Eigenvalue buckling analysis and its buckling factor

The transmission tower is a truss structure composed of a large number of components. When one component is subjected to instability deformation, the internal force characteristics of the whole structure will be affected and redistributed. Therefore, it is necessary to treat the whole structure as a whole and then conduct buckling analysis[7]. Considering the transmission tower as an ideal linear elastic structure, the theoretical buckling strength (i.e., the bifurcation point) is solved through the characteristic equation, and the calculation results are conservative, as shown in Fig. 1.

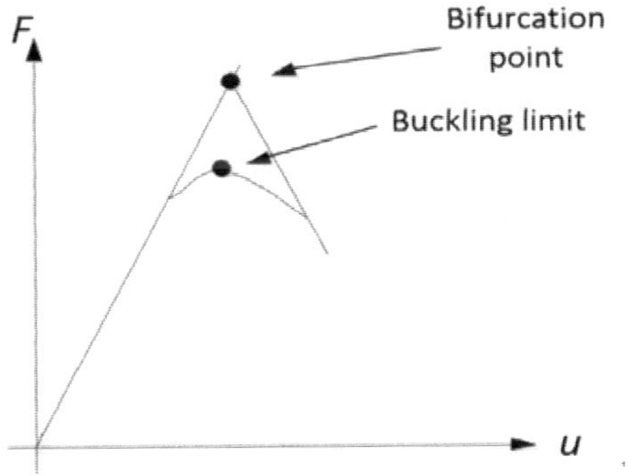

Figure 1. Eigenvalue buckling analysis

Under the condition of unit external load, let the structural stress stiffness matrix be $[K_\sigma]$, the load multiplier be λ, and the product of the two correspond to the geometric stiffness matrix of another strength.It and the usual stiffness matrix are not displacement functions. When the load matrix $[R]$ is constant, the displacement matrix $[D]$ under the baseline condition plus the virtual displacement $[\bar{D}]$ should satisfy the equilibrium state equation:

$$([K]+\lambda[K_\sigma])[D]=[R]$$
$$([K]+\lambda[K_\sigma])[D+\bar{D}]=[R] \quad (3)$$

We subtract this from that

$$([K]+\lambda[K_\sigma])[\bar{D}]=0 \quad (4)$$

The above equilibrium equation becomes unstable in the presence of the singular solution, let $([K]+\lambda[K_\sigma])=[K_{eq}]$, then:

$$\left|[K_{eq}]\right|<0;\ \textit{Unstable equilibrium state}$$
$$\left|[K_{eq}]\right|=0;\ \textit{Instability}$$
$$\left|[K_{eq}]\right|>0;\ \textit{stable state}$$

The critical load at the bifurcation point is the product of λ and the applied external load. When λ is negative, it means that the buckling load should be applied in the opposite direction.The displacement feature vector $[\bar{D}]$ represents the buckling mode and can easily show the buckling shape.

4. Transmission tower reinforcement

For the transmission tower that was put into operation earlier, the operation time of some components is too long and the performance declines, which leads to the insufficient stability of the tower structure. Therefore, the reinforcement measures for the rods have lower cost and can completely improve the stability and extend the safe operation time, with remarkable effect[8].

4.1. Composite section form

The new composite section form increases the sectional area, and its reinforcement effect is not completely equivalent to the method of increasing section.The cross section forms are mainly type X, type Z and type T, as shown in figure 2.The buckling analysis is carried out to study the reinforcement effect of the composite members.

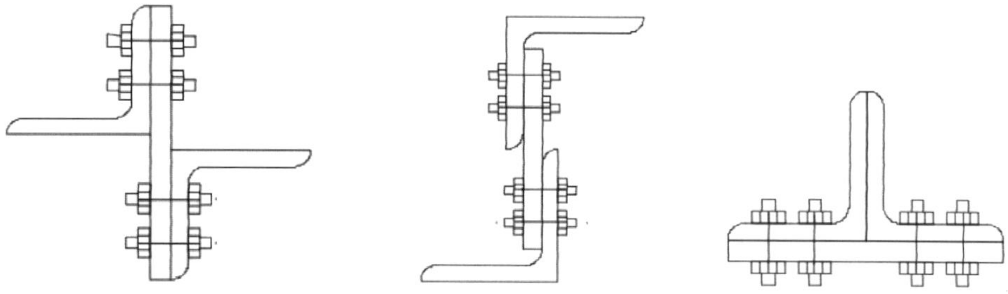

(a) X-shaped section (b) Z-shaped section (c) T-shaped section
Figure 2. Composite section form of reinforced component

5. Buckling analysis of transmission tower

IOP Conf. Series: Materials Science and Engineering **637** (2019) 012009 doi:10.1088/1757-899X/637/1/012009

The eigenvalue buckling analysis of the transmission tower is completed by ANSYS, and the main steps are as follows: 1) three-dimensional modeling;2) solve the static solution;3) eigenvalue buckling analysis and solution;4) expand the results;5) result list and drawing display.

5.1. Reinforcement condition and buckling solution

In this paper, three reinforcement schemes are modeled to analyze the effect of reinforcement and compare the situation without reinforcement.
(1) working condition 1: no reinforcement working condition.
(2) working condition 2: reinforce the first section of the tower.
(3) working condition 3: reinforce the first and second paragraphs of the tower.
In working condition 2 and working condition 3, cross section, z-shaped section and t-shaped section combination were adopted respectively, and the main material section was increased by 50%.According to the actual transmission tower structure modeling and applying constraints and wind loads in ANSYS software, the first 2 order buckling modes of eigenvalue buckling analysis are extracted. Figure 4 (a) and (b) are the corresponding results. The order and corresponding coefficients of buckling modes are shown in table 1.
Considering the actual situation of line operation and the situation that the wind speed continues to increase, the wind pressure of the tower increases accordingly, and the main material at the lower part of the tower body usually buckling first, causing the failure.As the weak link of the whole tower under strong wind, in order to improve its bearing capacity, this paper adopts the method of partial reinforcement of main materials and makes modeling analysis.See figure.3 for the segmentation of tower body.

5.2. Comparison of calculation results

The wind load was calculated according to the basic wind speed 30m/s and applied to the tower. The buckling modes of the tower under various working conditions were shown in figure 4. The analysis and comparison results were shown in table 1.

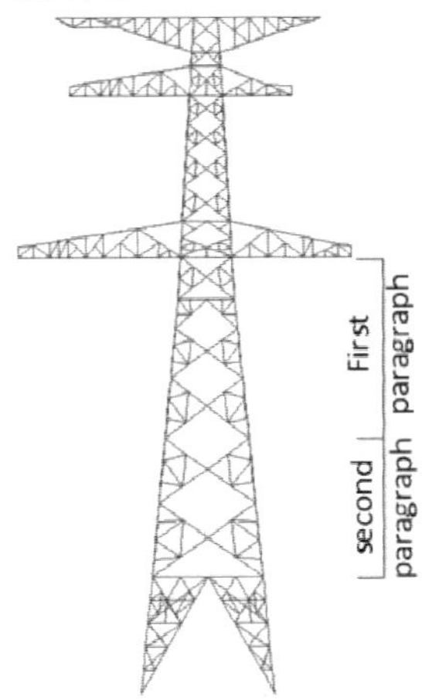

Figure 3. Schematic diagram of the body section

IOP Conf. Series: Materials Science and Engineering **637** (2019) 012009 doi:10.1088/1757-899X/637/1/012009

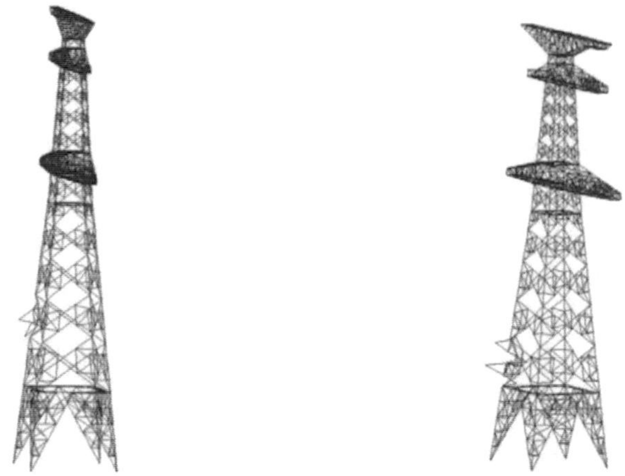

(a) First order buckling mode in case 1 (b) Second order buckling mode in case 1

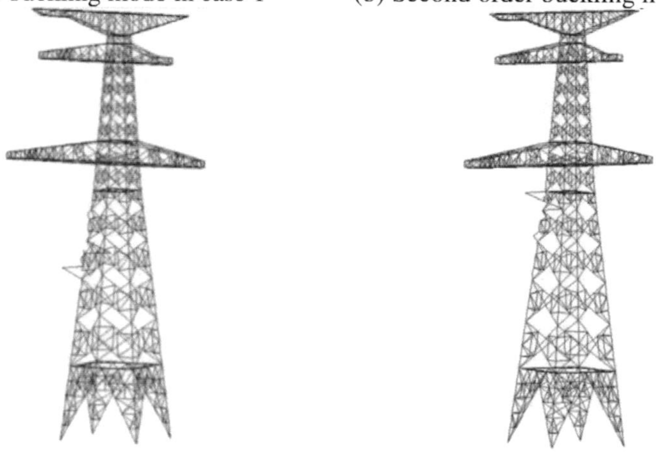

(c) First order buckling mode in case 2 (d) Second order buckling mode in case 2

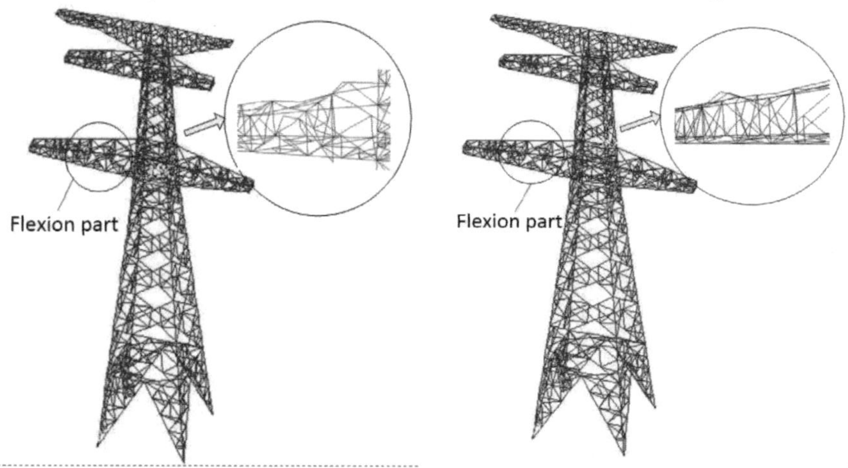

(e) First order buckling mode in case 3 (f) Second order buckling mode in case 3

Figure 4. The first and second buckling modes under different working conditions

Table 1. Mode and coefficient of wind resistance buckling of pole and towerin various working conditions

IOP Conf. Series: Materials Science and Engineering **637** (2019) 012009 doi:10.1088/1757-899X/637/1/012009

working condition	section form	order	buckling coefficient	The buckling coefficient increases by multiple	buckling mode
1	steel angle	1	7.218	——	Section 1 tower main timber instability
		2	7.306	——	
2	X-shaped section	1	8.165	1.13	Section 2 tower main timber instability
		2	8.840	1.21	
	Z-shaped section	1	9.079	1.26	Section 2 tower main timber instability
		2	9.830	1.35	
	T-shaped section	1	11.349	1.57	Section 2 tower main timber instability
		2	12.288	1.68	
3	X-shaped section	1	32.625	4.52	The first crossarm auxiliary is unstable
		2	35.507	4.86	
	Z-shaped section	1	36.261	5.02	The first crossarm auxiliary is unstable
		2	39.554	5.41	
	T-shaped section	1	45.218	6.26	The first crossarm auxiliary is unstable
		2	49.284	6.75	

It can be seen from figure 4 that under the wind load, the instability failure of the pole occurs at the lower part of the transmission tower. The results in table 3 show that the buckling coefficients of the x-shaped, z-shaped and t-shaped columns increase by 13%, 26% and 57% respectively after the completion of the reinforcement of the first section of the tower.After the reinforcement of the first and second sections of the tower body, the increase multiples of the buckling coefficients of the cross-shaped, z-shaped and t-shaped towers are 4.52, 5.02 and 6.26, respectively. When the main material at the lower end of transmission tower is strengthened, the weak link of buckling will transfer upward gradually, and the critical buckling load of the main material will increase accordingly.With the increase of buckling coefficient, the stability of the tower is improved. With the improvement of reinforcement degree, the buckling part is transferred to the crossarm auxiliary material part, at which time the buckling factor has exceeded 30 and the safety degree is relatively high. Moreover, the bars at the crossarm where buckling occurs account for about 16% of the whole crossarm, and the scope of buckling is small and the influence is weak. On the whole, after reasonable reinforcement, the transmission tower turns from the buckling failure of the main material to the instability failure of the partial auxiliary material of the transverse arm, and the overall stability of the structure is greatly improved, which can effectively avoid the collapse accident caused by the failure of the main material. In addition, the buckling coefficients of z-shaped and t-shaped increased by about 39% and 25%, respectively, under different x-shaped reinforcement schemes.The reinforcement scheme of t-shaped has better effect.

6. Conclusions

(1) by transforming wind load into transmission tower node load and applying it to ANSYS tower model by adjusting Angle steel orientation, the buckling modal analysis finds that the weak link of transmission tower instability is located at the lower part of tower body.

(2) the secondary loading scheme takes into account the actual stress of the tower before reinforcement in the actual project. The single point restart analysis technology including the

life and death units is adopted in ANSYS. Through the locking and activation of reinforcement units, the simulation can be effectively carried out to improve the accuracy of data.

(3) after the reinforcement of the main material of the transmission tower, the main material of the reinforcement part will not lose its stability, and the unstable part will move up along the main material.After the reinforcement of the first and second sections in the main timber is completed, the main timber will not lose its stability, and the unstable part will be moved to the crossarm auxiliary.

matrix.

7. References

[1] ZHANG Wen-liang, YU Yong-qing, SU Zhi-yi, et al. Investigation and analysis of icing and snowing disaster happened in Hunan power grid in 2008, J. Power System Technology, 32 (2008) 1-5.

[2] LI Cheng-rong, LU Yu-zhen, CUI Xiang, et al. Research issues for safe operation of power grid in China under ice-snow disasters, J. Power System Technology, 32 (2008) 14-22.

[3] LIU Xue-wu, XIA Kai-quan, GaoYan, et al. Study on strengthening the structure and its design for the transmission tower, J. Journal of Xian University of Architecture & Technology(Natural Science Edition), 43(2011) 838-844.

[4] LIU Hai-feng, ZHU Bin-rong, SUN Zhen-mao, et al. Experimental study on axial compressive strength and evaluation of design criteria for closely star-battened angle column with symmetrical battens of large width-to-thickness ratio, J. Journal of Building Structures, 39(2018) 123-130.

[5] LIU Hong-jun, LI Zheng-liang. Study of ultimate strength of dual-angle cross combined section of high-strength steel, J,. Engineering Mechanics, 30(2013) 140-146.

[6] LI Zheng-liang, HUANG Hai-bin, ZHAO Nan, et al. Study on mechanical performance of double internodes compression member of high-strength dual-angle cross combined section, J. Journal of Sichuan University (Engineering Science Edition), 47(2015) 64-74.

[7] GUO Yan-lin, WANG Xiao-an, ZHU Bo-li, et al. Experimental study on a four-angle assembled buckling-restrained brace subjected to axial compressive load, J. China Civil Engineering Journal, 50(2017) 1-12+115.

[8] LIU Shu-tang, XIONG Guo-wen, LIU Qi-hua, et al. Study on the strengthening of wind-resistant for transmission tower based on ANSYS, J. Journal of Guangzhou University (Natural Science Edition), 16(2017) 51-57.

IOP Conf. Series: Materials Science and Engineering **637** (2019) 012010 doi:10.1088/1757-899X/637/1/012010

Research on Mechanism of Film Pressure Loss in Water-lubricated Bearing

WANG Lun[1,2], WANG Nan[1,2*], DU Yubo[1], YANG Litao[1,2], WANG Peng[1,2], YUE Xiaokui[3]

[1]School of Mechanical Engineering, Shaanxi University of Technology, Hanzhong 723000, China

[2]Shaanxi Key Laboratory of Industrial Automation, Hanzhong 723000, China

[3]National Key Laboratory of Aerospace Flight Dynamics, Northwestern Poly technical University, Xi'an 710072, China

Email:1747776792@qq.com

Abstract. Due to the low viscosity of water, the water film is not easy to form, and the water-lubricated bearing system has the characteristics of airtightness. Therefore, it is difficult to obtain an accurate water film pressure distribution through experiments. Aiming at the loss of water film pressure in the process of water film pressure transmission caused by L-type pipeline in the existing wireless sensor testing methods for water film pressure, a mathematical model of water-lubricated bearing system pipeline pressure loss is established by numerical analysis method, and the corresponding physical model is established by using ANSYS software. The mechanism of pressure loss is studied by simulation. Finally, the test results were corrected and analyzed. The results show that the higher the flow rate of lubricating water at the inlet of the pipeline, the greater the pressure loss of the water film; the higher the shaft speed, the greater the local pressure loss; the longer the pipeline, the greater the pressure loss of the straight pipe section of the L-shaped pipeline, and the dynamic condition of the pipeline The pressure loss and energy loss are much more severe than static conditions.

1. Introduction

The water film pressure is the basic parameter of the water-lubricated bearing. The important characteristics of the bearing, such as the thickness of the water film, the lubrication state of the bearing, can be obtained through the in-depth study on the water-film pressure of the bearing. However, the water film pressure is not easy to obtain and the lubrication state of the bearing is more complex. The related research of water film pressure is usually based on numerical calculation or software simulation. Because mathematical and physical approximations are performed in the calculation[1~3] or simulation[4~6].Through the continuous efforts of researchers, several water film pressure test methods [7]have been put forward, but there are still different degrees of problems, the water film pressure at the limited point can only be measured, and the continuous pressure distribution of the radial section of the bearing cannot be obtained. In order to solve the above problems, after long-term research by our team, a non-invasive water lubrication bearing water film pressure is proposed, that is line sensing pressure test method. With the deepening of the research, it is gradually found that the proposed wireless sensing method of water film pressure, due to the installation of L-shaped diversion hole on the rotating shaft, will lead to pressure loss in the transmission process of water film pressure, which will affect the test accuracy. Therefore, based on the actual working conditions, the mathematical model of pressure loss of water lubrication system pipeline is established in this paper, and the shaft model and the physical model of pipeline are established by using ANSYS software. The influence of performance parameters on water film pressure of water-lubricated bearing under different working conditions is analyzed, and the basic law of pressure loss is obtained. Finally, the water film pressure test data of water-lubricated rubber bearings are corrected by using the research results, and the results are compared with the simulation results.

2. Problem Description

The water film pressure test bench is shown in reference[8]. Fig. 1 is the schematic diagram of the axial structure of the rotating shaft. Six radial draft holes are evenly spaced along the axis (R1~R6 is the water film pressure measurement point). At the right end of the shaft, there are six axial draft holes connected to the radial pilot holes (A1~A6 is the installation point of the sensor), and there are six axial pilot holes at the right end of the shaft (A1~A6 is the installation point of the sensor). The axial diversion hole is uniformly distributed in the circumferential direction and runs through the radial diversion hole. The water film pressure is transferred to the right sensor through the diversion hole for further testing.

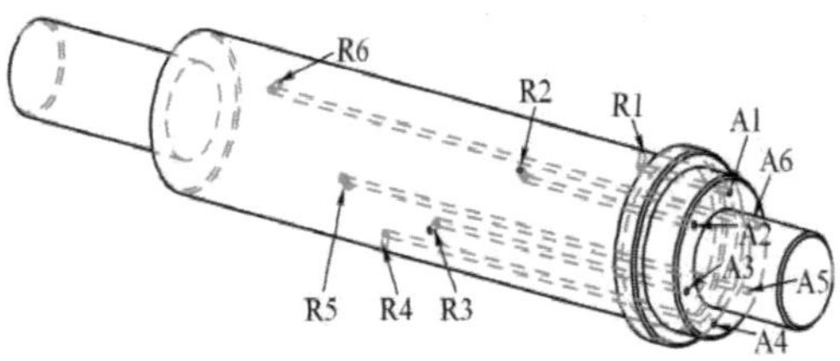

Figure 1. Diagram of rotating shaft

As can be seen from figure 1, due to the existence of L-shaped pipeline, the water film pressure will inevitably be affected when it is transmitted in the pipeline, and the pressure value obtained by the water film pressure sensor in the test process will also be different from the pressure value at the real measuring point. In this paper, the mathematical model of pressure loss of L-shaped pipeline is established by theoretical calculation method, and the physical model of pipeline is established by using CFD computational fluid dynamics module. The relationship between dynamic and static factors and pressure loss is analyzed, the law of pressure change is obtained, the mechanism of pressure loss is studied, Finally, the test results are corrected. Because the physical properties of the liquid in the test pipeline, such as viscosity, temperature and specific heat capacity, This things have little effect on the pressure loss of the pipeline, these factors are ignored in the numerical simulation. The flow chart of pipeline pressure loss mechanism in water lubricated bearing system is shown in Fig. 3. The methods are as follows: 1) to establish the theoretical mathematical model of the pipeline, 2) to establish the physical model of the pipeline, and to establish the physical model of the pipeline. 3) mesh the established model, set the boundary conditions, and select the relevant algorithm of fluid calculation; 4) The relevant output control attributes are set and relevant variables are set and relevant variables are set, and the parameters of the water lubrication bearing system pipeline are monitored and analyzed after the iteration is started to obtain the pressure loss-related mechanism.

3. Model establishment

3.1Mathematical model of pipeline pressure loss

According to the mechanism and manifestation of the energy loss in flow, the pressure loss can be divided into the pressure loss along the path and the local pressure loss. The pressure loss along the flow process refers to the flow loss caused by the viscous friction between the processes along the flow path. The mechanical energy consumed by the fluid to overcome the friction force along the flow process is called the energy loss along the way. According to the knowledge of fluid mechanics and Darcy's formula, the formula of pressure loss along the way can be written as shown in formula (1).

$$h_f = \lambda \frac{l}{d} \times \frac{v^2}{2g} \tag{1}$$

$$v = \frac{4q}{\pi d^2} \tag{2}$$

In formula (1), λ is the resistance coefficient along the way, which is related to the flow state (laminar flow or turbulence), Reynolds number and the roughness of the pipe wall, while for the L-shaped pipeline, assuming that it is a hydraulic smooth pipe, then λ can be obtained by formula (3).

$$\lambda = 0.3164\,\mathrm{Re}-0.25 \tag{3}$$

Joint formula(1), (2), (3) can be used to obtain pressure loss h_f:

$$h_f = \frac{8lq^2}{gd^5\pi^2} \times \lambda = (0.3164\,\mathrm{Re}-0.25) \times \frac{8lq^2}{gd^5\pi^2} \tag{4}$$

Local pressure loss refers to the flow loss caused by local obstacles in the process of fluid flow. For L-shaped pipe, there is local pressure loss at the bend. According to the knowledge of fluid mechanics, the calculation formula of local pressure loss is as follows:

$$h_j = \xi \frac{v^2}{2g} \tag{5}$$

The local resistance coefficients given in the existing literature are all measured under turbulent conditions. For 90°pipe, the calculation of the local drag coefficient under other turbulence conditions should be calculated according to the resistance coefficient along the path and equation.

$$\xi = \xi_0 \frac{\lambda}{0.022} \tag{6}$$

The simultaneous expressions （2）, (5), (6) can be used to derive the local pressure loss of the pipeline:

IOP Conf. Series: Materials Science and Engineering **637** (2019) 012010 doi:10.1088/1757-899X/637/1/012010

$$h_j = 419\frac{q^2}{\pi^2 d^4 g}(0.3164\,\mathrm{Re} - 0.25) \quad (7)$$

Among them, h_f、h_j - the pressure loss along the way;

l - pipe length;

q - the flow, unit in the pipeline;

d - pipe diameter;

v - liquid flow rate;

ξ - local resistance coefficient;

Δp - total pressure loss of the pipeline system, unit;

Re is the Reynolds number.

3.2 Pipeline physical model

When modeling, the structure such as fillet, chamfer and so on is ignored, and the calculation domain of pipe in the rotation axis is created. The hexahedral element grid is used to divide the single pipe (pipeline 6), and the tetrahedral unstructured grid is used to divide the whole computing domain. As shown in figure 2, the number of nodes and units in a single pipe grid is 552883 and 578048 respectively. After the rotation axis is Boolean, the number of nodes in the computational fluid network is 110536 and the number of cells is 562531. Because the temperature change of water flows in the bearing system is small, the influence of temperature on the pressure loss of water lubricated bearing system is not taken into account.

（a） Exit grid （b） Local grid

Figure 2. Single pipe meshing

The properties of the fluid in the pipe are set as follows: the temperature is 298K, density is998kg/m^3 , the dynamic viscosity is $0.001 Pa \cdot s$, the inlet boundary is set as the velocity inlet, the turbulence intensity is set as 5, the hydraulic diameter is set as 4, the outlet condition is set as the free flow outlet, and the pipeline wall in the water lubricated bearing system is set as the non-slip wall. At the same time, the rotating speed of the pipe is set, and the different rotating speed is given. In water lubricated bearing system, the radius of shaft diameter is 50mm, and The gap ratio is c/R=0.008, The change range of speed spindle w is 47~200Rad/s,from the Taylor criterion,the fluid is turbulent. In the water lubricated bearing system, the fluid motion in the pipeline and the rotation of the pipeline itself are from the practical engineering point of view, so the turbulence model is selected as the standard turbulence model RNG $k-\varepsilon$, select water-liquid in material set up , density is998kg/m^3 , Viscosity is0.001Pa.s,thepressure-velocity couping algorithm is selected to solve the SIMPLE algorithm.

4. Simulation results and analysis

4.1 Effect of different rotating Speed on pressure loss of Pipeline

In order to explore the effect of different rotating speed on the pressure loss of water film, the longest pipe 6 was taken as the object. By changing the rotation speed of the shaft, it means that the speed of the pipe is84rad/s、126rad/s, respectively, the influence of different rotating speed on the distribution of pipeline pressure loss is analyzed. The results are as follows:

(a)Static pressure cloud diagram at 84rad/s

（b）Static pressure cloud diagram at 84rad/s

Figure 3. Pressure cloud map at different speeds

It can be seen from figure 3 that when the liquid in the pipeline flows at different rotating speed and the same speed, the pressure value at the entrance of the pipeline is the maximum, and the pressure of the radial pipeline decreases evenly along the radial direction, and there is no obvious sudden change. At the turning point of the L pipeline, due to the rotation of the pipeline and the fluid itself, the pipeline is affected by gravity, flow resistance, centrifugal force, coriander force caused by rotation, and the direction of velocity changes instantly in the local area, as a result, the fluid is gradually dumped near the outer wall, resulting in a large number of fluids being squeezed to the outer wall, the lateral pressure is higher than the inner pressure, and a vortex of different sizes is formed in the local area, and the vortex is produced. It is easy to be subjected to hydraulic impact, which leads to relatively large pressure loss and energy loss in the local area, resulting in changes in fluid viscosity, temperature and other properties. After the fluid reaches the horizontal pipe section, it advances in a stable form. As the distance between the local part of the pipeline and the exit is farther, the pressure loss value in the pipeline will be reduced. At the outlet of the pipeline, the pressure value in the pipeline is negative. The phenomenon of reflux shows that the pressure in the pipeline is continuous. At the same time, the decrease of pressure in the pipeline increases the resistance of liquid flow and promotes the liquid in the pipeline to flow forward.

4.2 Influence of different flow rate on pressure loss of Pipeline

The flow rate of the pump is 3.5m^3/h. The velocity of liquid in water lubricated bearing is the ratio of the flow value of pump to the cross section of pipeline inlet. In order to study the influence of different inlet flow value (velocity value) on the pressure loss of water film, By changing the velocity value at the entrance, that is the pipeline velocity values are 20m/s、25m/s respectively, the influence of different flow rate (inlet velocity) on the pressure loss distribution of the pipeline is analyzed, and the results are shown in figure 4.

(a) Static pressure cloud at a speed of 20m/s

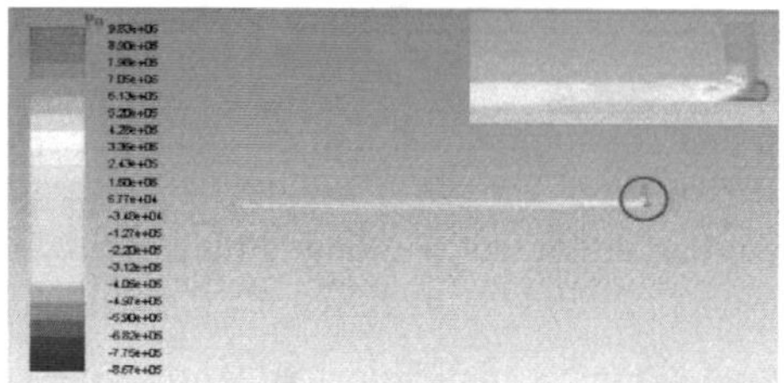

(b) Static pressure cloud at a speed of 25m/s

Figure 4. Static pressure cloud at different speeds

As can be seen from figure 4, the greater the flow rate (inlet speed) in the pipeline, the more the pressure decreases along the way. The fluid in the pipeline is always subjected to gravity, flow resistance, centrifugal force, and the Coriolis force produced by the pipe during rotation. When the liquid just enters the pipeline, the centrifugal force in the pipe is greater than the Coriolis force, resulting in a drop in the pressure in the pipeline. In the local part, the liquid in the pipe is caused by the double action of self-rotation and pipe rotation, which leads to the inconsistency of the internal and external pressure, resulting in the vortex, which is easy to cause vibration and noise in the local part. The horizontal straight pipe section, due to the rotation of the pipe, the direction of centrifugal force, the direction of flow resistance, and the direction of velocity, the change leads to the fluctuation of the pressure value in the pipeline. The larger the velocity of flow, the smaller the resistance, the smaller the velocity of liquid flow, the greater the resistance, the more pressure loss is caused in the horizontal straight pipe than the local pressure loss, and the higher the inlet flow rate is, the more the pressure loss is caused by the horizontal straight pipe, and the higher the inlet flow rate is. The larger the pressure value at the outlet of the pipeline is, the more the reflux occurs at the outlet, and the loss of the pressure value at the outlet increases accordingly.

4.3 Effect of different sizes on pressure loss of Pipeline

In order to study the influence of six pipelines of different sizes on the pressure loss of the pipeline, the idea of control variable method is used to analyze the pressure loss of the pipeline. The reference conditions of the parameter setting are as follows: the rotational speed is 480r/min, the pressure of water supply is 0.25MPa ,The pressure based solver is selected, the transient simulation is selected, and the RNG is selected. In the turbulence model, the velocity inlet is selected, the turbulence intensity is 5, and the velocity at the entrance is 5m/s the rated velocity at the test. The temperature is 298K , hydraulic diameter is 4mm , the exit is set to the free flow exit, and the rotating area is set to rotate at the speed of 480r/min (50rad/s) , the center of rotation is (0, 0, 0), and the direction of rotation is clockwise around the Z axis The wall area is set to a non-slip wall, and the absolute velocity is set to 5m/s, the speed is set to 50m/s ,The SIMPLE algorithm is selected and the second order upwind scheme is selected.

At the rotating speed of 480r / min , the pressure of water supply is 0.25MPa, not in the external load, The flow rate at the rated entrance is 5m/s . The pipe pressure cloud diagram of different sizes is shown in figure 5 (a). From figure 5 (a), figure 5 (b), it can be seen that the influence values of different sizes on the pressure loss of the pipeline are different without external load. The longer the pipeline, pipe pressure loss of the greater value, in the local bend pipe, the pipe pressure value decreased trend of local area is larger, and the lateral pressure is greater than the inside of the pressure value in local place, because in the local place the direction and the rotation of the rotor speed in pipeline direction constantly change, caused by the changing of fluid pressure value in the pipe. It is calculated that the pressure loss value of outlet pipe 1 is $4.5KPa$.Pipe 2 pressure loss value is $1.870KPa$.Tube 3 pressure loss value is $4.30KPa$.The pressure loss value of pipe 4 is $5.23KPa$.The pressure loss value of pipe 5 is $6.037KPa$. The pressure loss value of pipe 6 is $7.25KPa$.

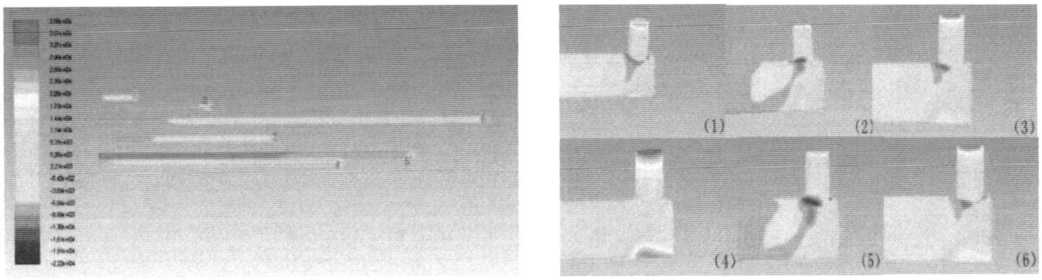

(a)Overall static pressure loss (b) Local pressure changes at different sizes of pipes

Figure 5. Pressure loss cloud map of different sizes

5. Test result correction

In the case of simulation, from the simulation of the whole pipeline, we can see that the pressure value in the pipeline begins to drop from the entrance, and then the pressure value in the pipeline begins to have reflux in the horizontal stage, and the reflux pressure is larger in the vertical. The opposite changes of vertical pipe and horizontal pipeline in different parts ensure that the water-lubricated bearing system tends to a dynamic stable process in the case of rotation. The mathematical statistical method is used to calculate the ratio between the entrance and outlet sections of pipeline 4 and pipeline 6, respectively, and then the ratio between inlet and outlet is calculated by means of mean value. The specific methods are as follows: the average pressure on the surface at the section of pipe 4 and pipe 6 is obtained by using the area setting panel (Surface Integrals) in node Reports of the model tree, and the report type (Report Type) is selected as the surface average pressure (Area-weighted Average). The inlet (in4) and outlet (out4) of pipe 4, the inlet (in6) and exit (out6) of tube 6 are selected respectively, and the average pressure values of inlet and outlet sections are obtained, respectively., and the report type (Report Type) is selected as the surface average pressure (Area-weighted Average),). The inlet (in4) and outlet (out4) of pipe 4, the inlet (in6) and outlet (out6) of pipe 6 are selected respectively, and the average pressure values of inlet and outlet sections are obtained respectively. By calculation, the proportional coefficient of outlet and entrance of pipe 4 is 0.63, which is the same as that of pipe 4. Use the same approach to manage By calculating the coefficient of channel 6, it can be concluded that the ratio of pressure coefficient between outlet and inlet of pipeline 6 is 0.711, and then the pressure value at the point measured by the real sensor with pressure coefficient is drawn, That is, the test results of the full cycle distribution of water film pressure in the radial section of rubber bearings with eight grooves and concave surfaces at the real measuring point (in the case of pressure loss), the pressure unit is kPa, load unit is N. Figure 6(a), figure 6(b) is the true real value of pipeline 4 and pipeline 6, respectively.

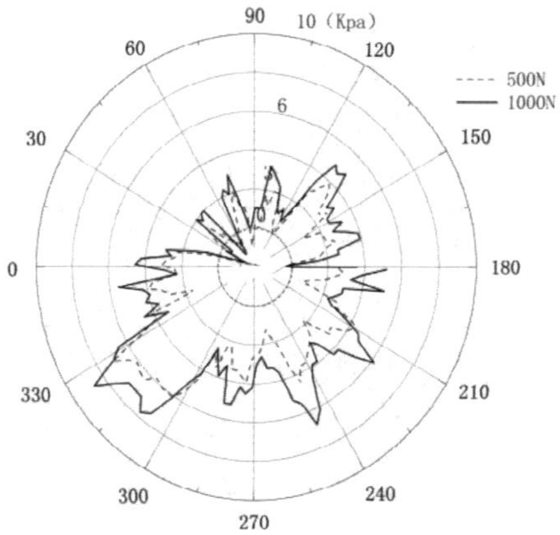

(a) Water film pressure distribution over the entire circumference of tube 4 after calibretion

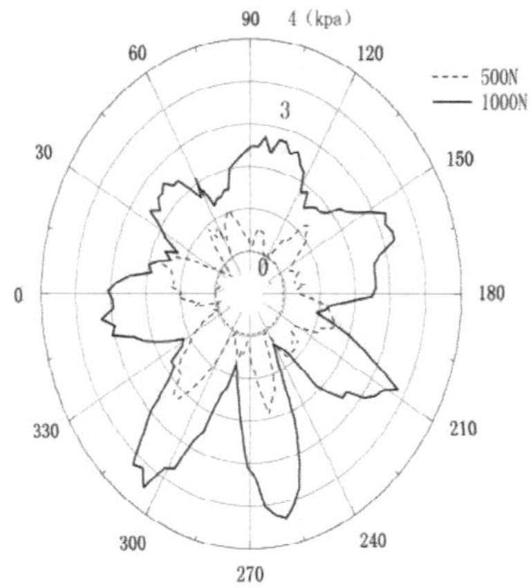

(b) Water film pressure distribution over the entire circumference of tube 6 after calibretion

Figure 6. Water film pressure distribution over the entire circumference of tube 4 and 6 after calibretion

From figure 6(a) and 6(b), it can be found that there is still a difference between the simulation and the test, and there are some errors. When the load is 500N and 1000N, and the shaft speed is 480r/min, water supply pressure is 0.25MPa; the water film pressure in the pipeline is redistributed, the pressure in the pipeline changes regularly along the axial direction. Through the experimental correction results, it can be seen that there is a large pressure between each strip and the groove, and the reason for this change is that there is a pressure loss in the pipeline. There is a large and small water film between the laths (shown in the concave area of the figure), that is, the pressure value on the plate appears in the center of the lath. From the results of figure 6(a) and figure 6(b) after modifying the water film pressure, it can be seen that the water film pressure in the pipeline also forms a pressure peak, which makes different parts of the lath bulge and forms a water capsule, resulting in the formation of fluid pressure. However, the pressure value in the pipeline is in a kind of dynamic change, and the pipeline reflux phenomenon appears in the horizontal part of the pipeline, but at this time, the pressure value in the pipeline is larger in the vertical. It is shown that the dynamic change process leads to the existence of pressure difference in the pipeline, resulting in the energy loss in the pipeline and the pressure loss in the pipeline.

6. Conclusions

In order to solve the problem of pressure loss in wireless monitoring of water film pressure of water-lubricated bearings, the mathematical model and physical model of pipeline pressure loss are established, the mechanism of water film pressure loss of L-shaped pipeline is studied, and the test results are corrected. The conclusions are as follows:

（1）The factors affecting the pressure loss of water film are the flow rate of liquid, the rotating speed of rotating shaft, the length of diversion hole in the design of rotating shaft, and the dynamic and static state of the rotating shaft. Because of the pressure loss of the rotating shaft in the process of rotation, it is easy to have the pressure difference at the corner. The axial water film pressure shows a downward trend, and it fluctuates obviously in the middle of the rotating shaft. This is because there is pressure loss in the rotating process of the rotating shaft, at the same time, the internal and external pressure difference is easy to occur at the corner of the rotating shaft, and the vibration and noise of the pipeline are easy to produce. Secondly, there will be liquid reflux near the monitoring point of the sensor.

（2）The speed, velocity and size of the pipeline will cause the pressure loss in the pipeline, and the larger these parameters are, the greater the pressure loss will be, and the pressure drop value in the pipeline will maintain a dynamic change process.

7. References

[1] LIU Gang, Li Ming. Analysis on Lubrication Characteristics of Water-Lubricated Rubber Bearings Under Wear Condition[J]. Bearing.2018(08):44-48.

[2] FANG Bin. Preliminary research on calculation for stiff ness of marine water-lubricated bearing[J]. Ship.2017,28(04):69-73.

[3] JIANG Hongyan, CHEN Jiajun, YAO Jian et al.Static and Dynamic Characteristics of Water-lubricated Hydrodynamic Bearings Considering Nano-bubbles Effects[J]..Lubrication Engineering. 2017,42(01):84-90+96.

[4] Ye Xiaoyan, Feng Yaoning, Wang Jing， et al.Test and analysis of water-lubricated bearing of seawater desalination pump[J]. Journal of Drainage and Irrigation Machinery.2015,33（06）：510-515.

[5] YANG Sen, LI Ming.Elastohydrodynamic Lubrication Characteristics of Water-Lubricated Rubber Bearings with Mixed Groove Structure[J].Bearing.2018（02）:38-42.

[6] LIU Jialei, LI Ming, LIU Gang.Two-way Thermal-Fluid-Solid Interaction Lubrication Characteristic Analysis of Water-lubricated Rubber Bearings[J]..Lubrication Engineering.2018,43（01）:81-86.

[7] Litwin W. Influence of surface roughness topography on properties of water-lubricated polymer bearings[J]. Tribology Transactions, 2011, 54(3):351-361.

[8] WANG Nan ,MENG Qingfeng, WANG Pengpeng, et al. Simulation and Experimental Research on Properties of Water-lubricated Rubber Bearing[J].Journal of Mechanical Engineering,2014,50（13）:113-121.

Acknowledgements

This work was financially supported by the National Natural Science Foundation of China (51605269) and the Young Outstanding Talent Support Program of University in Shaanxi Province (SLGQD1802, SLGD1812).

Application of new automatic welding equipment and welding technology in the future construction machinery field

ZHANG Yan, ZHANG Fei

Huanghebei Road, Xiyanduan district, Xuzhou City, Jiangsu Province, China
Email: zyredeemer@outlook.com

Abstract. The production efficiency of machinery manufacturing industry is affected by the automatic welding equipments and technologies directly. Through introducing the working principle and applicable working conditions of new automatic welding equipments and welding technologies, the writer uses laser tracking equipment and welding deformation technology to analyze the possibility of application of automatic welding technology in the field of construction machinery, so as to provide useful reference for improving the quality of mechanical welding, reducing production costs and strengthening brand effect.

Keywords:automatic welding, laser tracking, double wire triple arc welding, cold arc welding, fish scale welding

1. Introduction

The global manufacturing industry is declining nowadays. The PMI index of the global manufacturing industry has continued to decline by the end of 2018. The market stock of the construction machinery industry is far larger than the market demanding, and it is facing unprecedented challenges. Looking at the whole market, the total construction machinery manufacturing industry has entered a state of poor operation. The shrinkage of the production of the world's major well-known construction machinery industry and the corresponding measures that they are taking indicate that the future development of the industry will still be in a cold winter. Facing with such a severe market situation, improving product quality, reducing production costs and using costs, and widening the gap between brand effects are the keys to win the battle. After more than two centuries of continuous development in the construction machinery industry, the threshold of traditional manufacturing technology has been greatly lowered, and production automation, lightweight products and energy-saving have gradually become the focus of future competition.

With the development of industrial robotics technology, it has been widely used in machinery manufacturing industry, such as cutting, welding, painting, assembly, etc. It improves the product quality and production efficiency of manufacturing industry significantly, and the technology is matured day after day, which undoubtedly raises the new height of construction machinery manufacturing industry. However, Industrial robots also has its disadvantages. Let's take welding robot as example, its material is expensive, lamella piece can't be tracked by arc accurately, easy deformation structure cannot be welded accurately by using multilayer multichannel method, thin wall pipe joint cannot be welded, and fusion at the root of narrow groove is not so good. The writer solves the above technical bottleneck with the experience which was learned in the practice of the automatic welding equipment and welding technology.

2. Application of Laser Tracking Equipment

The key of robot automatic welding is intelligent positioning and arc tracking technology. At present, most intelligent positioning technologies of welding robots mostly use welding wire contact method, and some welding robots choose nozzle contact positioning technology. For example, IGM welding robot. With the application of laser tracking equipment, the deficiency of intelligent positioning and arc tracking technology is effectively supplemented.
used if desired. If an author has additional information to appear as a footnote, such as a permanent address or to indicate that they are the corresponding author, the footnote should be entered after the surname.

2.1. Searching Principle of Welding and Its Defects

The robot contact positioning circuit is a constant open loop. When the welding wire or nozzle contacts the workpiece, the constant open loop closes. The robot built-in relay closes. The processor records the position information of the contact point. In the positioning process, the deviation between the actual position of the reference point on the workpiece surface and the programming position can be determined, and the corresponding welding procedure can be corrected.
When the errors of welding positioning occur, the position of the weld seam will change with the robotic automatic correction, which is the most fundamental reason for welding defects in robotic automatic welding. Taking contact positioning of welding wire as example, in order to ensure the positioning accuracy, it is necessary to control the dry elongation of the welding wire when positioning. At this moment, the role of wire clamping gun cable is very important. With the aging of the clamping gun cable and the blocking of the clamping cylinder, the reliability of the clamping wire decreases gradually. In the case of sheet metal welding below 3mm or back welding with face positioning, the change of dry elongation will easily lead to the deviation of positioning, which will lead to welding defects. In addition, there are many factors which affect the contact positioning accuracy of welding wire, such as damage of conductive nozzle, surface quality of workpiece, conductivity of welding wire and so on. Similarly, the contact positioning of the nozzle also has its disadvantages, such as the nozzle is easy to be deformed by heat, the nozzle is easy to be adhered with spatter, and the conductivity of the nozzle is reduced.

2.2. Arc tracking technology and its defects

At present, the seam tracking technology used in automatic welding on the market is arc tracking technology. In the process of fillet welding or thick plate welding, the welding gun generally needs to swing horizontally. Because of the different dry elongation of the welding wire, the actual welding current is different from the set current. The shorter the dry elongation of the welding wire, the bigger the actual welding current, and the longer the dry elongation of the welding wire, the smaller the actual welding current. Using this principle, the welding robot can process the detected electrorheological changes and the position of the welding torch in real time by corresponding software, and then correct the actual welding trajectory of the robot, so as to ensure that the center line of the welding trajectory is always in the middle of the groove, and at the same time ensure the consistency of the height direction of the welding torch and the welding seam [1].

Among many droplet transition forms, jet transition welding is the most ideal arc tracking state which has stable process, less spatter, large depth and perfect welding seam[2]. In order to ensure a good arc tracking state, the actual welding current is greater than the minimum current from the droplet to the jet during the welding process. Taking 1.2mm ER50-6 wire mixed gas shielded welding as an example, its welding critical current is 240A, when the welding current is lower than 240A, arc tracking accuracy tends to be unstable. So the arc tracing technology is suitable for welding of high current and medium thickness plate structure. When the thickness of the plate is less than 5mm, the arc tracking accuracy begins to decline, and when the thickness of the plate continues to drop below 3mm, the arc tracking function fails. Another important factor that makes thin sheet metal unsuitable for arc tracing technology is that it is easy to overheat the edge with smaller lap edge during welding. The higher the temperature, the larger the resistance and the smaller the welding current. When the arc tracking is used, the gun will automatically correct to the direction of the smaller current, which will lead to the deviation of the weld bead and the melting of the base metal with short lap edge. In addition, the surface quality of weld, joint quality, chemical composition of welding wire and mixed gas will affect the arc tracking effect.

2.3. Laser tracking principle and characteristics

Laser tracking is a technology that laser irradiates a moving target and controls the direction of the measurement system according to the deviation angle between the laser signal reflected from the target and the optical axis of the measurement system. Laser tracking system usually consists of laser illuminator, laser detector, information processing system and servo system. The laser signal emitted by the laser illuminator to the target is reflected back by the target and projected onto the detector through the optical system. Laser detectors are generally four-quadrant photodetectors. If the signal reflected back from the target has a deviation angle from the optical axis of the measurement system, the area of the spot projected on the photodetector in the four quadrants is not equal. The corresponding angular deviation signal is obtained after the information is processed. This signal drives the servo system to make the measurement system turn to the target until the deviation angle is zero, thus it can complete the process of laser tracking [3].

The laser tracking device developed at present can not only realize efficient seam tracking, but also realize intelligent seam positioning. Laser positioning can not only make use of laser ranging technology for contact positioning of weld seam, but also use laser imaging technology to measure the height and transverse position deviation of weld seam at one time. The non-working time of robot can be reduced by 5-8 times [4]. Therefore, the application of laser tracking technology can effectively solve the defects of welding positioning and arc tracking technology.

Laser positioning and laser tracking technology can solve the problems of variation of dry elongation

IOP Conf. Series: Materials Science and Engineering **637** (2019) 012011 doi:10.1088/1757-899X/637/1/012011

and fluctuation of welding current, but higher quality of workpiece surface and joint precision which it requires are high. Rough weld surface or fluctuating joint gap will generate diffuse reflection in laser imaging and reduces the laser tracking precision. Because of the relatively large volume and high cost of the laser tracker, it can be generally used as a supplement to the contact positioning and arc tracking technology and applied to the welding of large open structural parts or thin plates.

3. Welding Deformation Control Technology

During welding, the melted metal and the base metal near the seam expand by heating, thus it results plastic deformation. When solidified, the metal which is in the welding part and near the welding seam shrinks, this will result internal stress in longitudinal and transverse, which makes the longitudinal and transverse shrinkage of the weld part and forms the deformation. As cold shrinkage and thermal expansion are inevitable during welding, so the welding deformation is inevitable. We can only control but not eliminate welding deformation completely. There are many factors affecting welding deformation, such as the features of base metal, form of weld seam, welding heat input, welding direction, joint clearance, deposited metal amount, etc. Under certain welding environment, reducing welding heat input and optimizing welding sequence are undoubtedly the most effective methods to control welding deformation. The following introduction of welding technology provides a new direction for reducing welding deformation.

3.1. Cold Arc Welding Technology

with the advancement of industry, the application of thin sheet and light metal structure parts is more and more extensive. In order to reduce plate welding deformation and avoid plate overheating breakdown, the welding heat input of this kind of plate must be controlled. In the process of welding, short circuit transition welding with low current and voltage is the best choice for traditional thin plate welding [5].The droplet contacts with the molten pool before it disconnects from the electrode end directly. The arc is put out and short-circuited. The droplet transfers to the molten pool under the action of electromagnetic contraction force generated by the short-circuit current and surface tension of the liquid metal. At the moment of short circuit, the voltage drops sharply and the current rises gradually as the resistance decreases. At the moment of droplet falling off, the voltage rises sharply to ensure the arc reburning, while the current can only slow down under the action of inductance. At this moment, the current and voltage both are high, and welding defects such as arc instability, undercut, welding spatter and breakdown are easy to appear.

Cold arc welding technology is to use digital signal processor to quickly adjust the current and voltage according to feedback at the moment of arc reburning, and to reduce the welding current at the moment of arc reburning, so as to ensure the stability of arc reburning. After arc initiation, a small pulse current is given to make the arc obtain higher arc energy and promote the re-nucleation of melt droplets. The cold arc welding technology makes the melting droplet transition period stable, guarantees low heat input, and greatly reduces the welding spatter caused by traditional short-circuit transition.

3.2. Technology of Double-wire Three Arc Welding

The welding parts in the construction machinery industry are mainly medium and thick plates. Double-wire automatic welding can effectively improve the welding efficiency. It is widely used in the welding of key seams such as loading mobile arms, big arms of excavator and crane. In addition to increasing welding efficiency, welding heat input also increases greatly. Thermal deformation after welding has always been a problem troubling the construction machinery industry.

Different from traditional double-wire welding, double-wire three arc welding can form a third arc between double wire arcs. In the double-wire three arc welding system, the arc power used to melt the wire is much larger than that used to melt the base metal. The melting efficiency is greatly improved, and the welding heat input is significantly reduced. The arc power between three arcs can also be allocated by intelligent control of digital signals. When higher melting depth is needed, the arc power for melting base metal can be increased, the arc power for melting wire can be reduced, and enough heat input can be guaranteed for melting the base metal. Similarly, when it is necessary to control the welding deformation and improve the welding efficiency, the arc power used to melt the base metal can be reduced and the arc power used to melt the welding wire can be increased to ensure sufficient melting efficiency.

3.3 Fish Scale Welding Technology

The difficulty of thin-wall pipe welding lies in the control of welding posture, especially the welding of joints of through-line pipe wall. Traditionally, tungsten argon arc welding, brazing or short-circuit transition gas shielded welding are often used in the welding of such welding seams. Among them, the cost of tungsten argon arc welding is high, the welding strength of brazing is difficult to guarantee, and gas shielded welding is difficult to control the heat of welding when vertical welding and cross welding are used, which is easy to break down the pipe wall.

The introduction of fish scale welding technology provides a possibility for automatic welding of bridge rail and tower crane support frame. During the whole process of fish scale welding, the robot always makes a step forward movement. For each distance the robot goes ahead, it will stay in the same place for a while, and then steps forward gain, and so on until the end of welding. When the robot stops, the welding power source starts arc welding rapidly. When the robot moves, the welding machine switches off instantly. The ratio of pulse of fish scale welding is 35%~55%(generally 40%) per second. Compared with the traditional welding process, the whole welding heat input is greatly reduced, all position welding can be realized, and the welding seam is flaky fish scale. At present, it has been widely used in welding bicycle and motorcycle frame, fitness equipment and medical equipment.

4. Concluding remarks

The welding equipment and technology outlined by the writer is only the corner of the iceberg of many new welding technologies. Welding automation technology is a comprehensive subject which integrates welding, automation, machinery, hydraulic pressure and measurement. It requires stable preparation before welding, skilled process monitoring and professional performance maintenance. Even if the advanced welding technology is available, the ideal welding quality cannot be achieved without proper structure, mature welding technology, stable joint quality and excellent personnel quality. In increasingly fierce competition of today, the construction machinery industry should consolidate the traditional essence internally and study the advanced technology from external, upgrade the new manufacturing level with advanced production technology continuously.

IOP Conf. Series: Materials Science and Engineering **637** (2019) 012011 doi:10.1088/1757-899X/637/1/012011

References:

[1] Qing Weifu, Luo Bin. Application of Arc Sensing Technology in Weld Tracking,J. *Metal Processing (Hot Processing)*, (2014)40-41.

[2] Zhang Zheng. Research on Key Technologies of Robot Welding for Large Structural Parts of Ships ,D. *Full-text Database of Excellent Master's Degree Thesis of China*, (2010).

[3] Information on www.baidu.com.2019.

[4] Lin Laixing. A Sensor for Measuring the Whole Process of Rendezvous and Docking: Composite LIDAR,J. *Space Control*, (1992)41-48.

[5] Zhang Hong. EWM cold Arc "Cold Arc Welding" - A Low Energy Melting Pole Gas Shielded Welding Technology and Its Application,J. *Welding*,(2009)13-14.

Practical Research on Welding Simulation of Q345 Steel

Zhi-ling Wang

Jiangsu Province Xuzhou Technician Institute, Xuzhou 221151, China
379713268@qq.com

Abstract. In this paper, we use TCL/TK integrated language analysis software and semi-ellipsoid heat source model and life-death element method to simulate the butt welding process of Q345 steel plate with thickness of 10 mm. We simulated and analyzed the temperature field, stress field and welding deformation of multi-layer welding. Through welding simulation analysis, we extracted the distribution of welding residual stress and welding deformation of butt joint, and obtained the transverse and longitudinal shrinkage of butt plate and the main reasons. The results show that the multi-layer butt welding seam and its adjacent area are under tension, and the numerical value generally reaches the yield strength, and the two sides are under compression. The tensile stress in the middle of the weld bead is the largest and decreases gradually toward both ends. The cross-sectional area of single-sided multi-layer welding seam is larger and asymmetric along the thickness of the plate. After welding, the transverse shrinkage and angular deformation of the welded parts are larger. At the same time, the accuracy of the simulation results is verified by actual welding and testing.

1. Introduction

As an important method for solving complex engineering problems, finite element method is widely used [1]. With the development of computer technology, more and more scientists and technicians are engaged in the research of process simulation in order to optimize design and process parameters, reduce test cost, improve production efficiency and product quality. In the field of welding, many progresses have been made to optimize the process by numerical simulation [2]. But in the pretreatment process of the model, each weld must correspond to a working condition. At the same time, the welding path, filling and boundary conditions should be set. When welding the whole structure, the efficiency is very low. In engineering applications, various kinds of special finite element software have their own characteristics in geometric modeling, meshing, analysis and calculation. Although in many cases only one software is needed to complete the analysis process of the whole model, the pretreatment of the model is complex, inefficient and error-prone [3].In this paper, we use TCL/TK software resources to establish the software integration environment, and simulate the welding process and method of Q345.We simulated the temperature field, stress field and welding deformation of multi-layer welding and compared them. After extracting the distribution of welding residual stress and welding deformation of butt joint, we verified the main reason of transverse shrinkage of butt plate.

2. Welding simulation analysis process

The procedure of welding simulation analysis using thermo-elastic-plastic finite element method is shown in Figure 1. Firstly, the geometric model is transformed into mesh model, then the welding heat source parameters and material parameters are input, and the thermal boundary conditions are applied to calculate the welding temperature field. Finally, we take the temperature field and structural constraints as the boundary conditions to analyze the welding deformation and residual stress, and get the results of welding simulation analysis in the analysis of welding structure [4].

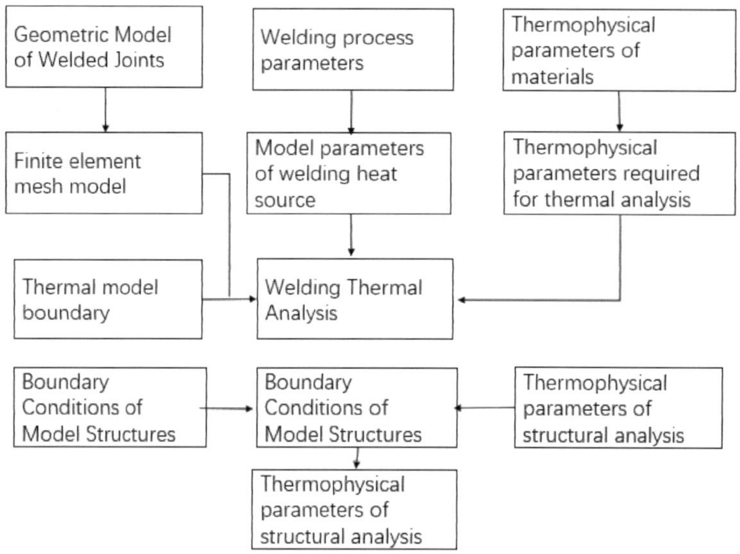

Figure 1. Analysis process of welding simulation

3. welding simulation example

3.1. Physical Model Establishment.

Firstly, we prepare two Q345 steel plates with a dimension of 200 mm x 200 mm x 10 mm and a V-groove. The geometric model of the steel plate is shown in Figure 2. Next, we use MAG method to weld, and the welding parameters are shown in Table1.

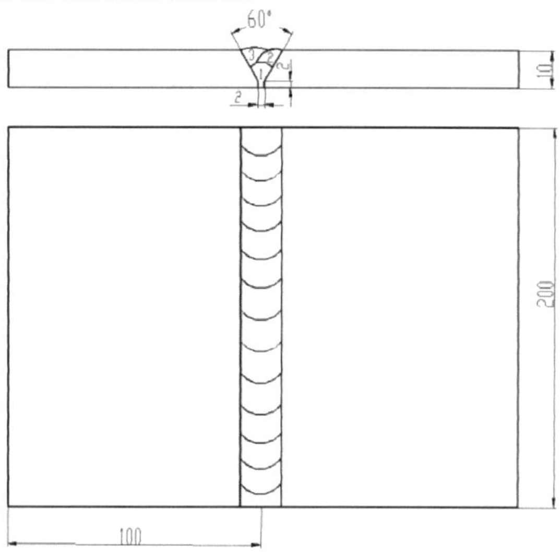

Figure 2. Geometrical model of butt plate

Table 1. Welding process parameters

Number of welding layers	Welding current I/A	Welding voltage U/V	Welding speed V/(mm·s-1)
First floor	240	28	5
Second floor	295	32	7
Third floor	295	32	7

3.2. Establishment of Finite Element Model.

3.2.1. Finite Element Parameter Setting. When establishing the simulation model file of welding simulation, we divide the solid element mesh into three layers to ensure the accuracy of calculation. At the same time, in order to improve the calculation speed, we control the element mesh of the weld and heat-affected zone to 2 mm, while the element mesh far from the weld zone to 6 mm [6]. According to the characteristics of MAG welding and the restraint mode on site, we choose double ellipsoid heat source as the boundary condition of welding heat source, set the heat transfer coefficient to 0.02N/mm2/sec/K, define the mechanical boundary condition with displacement restraint, and set the heat source type, heat dissipation coefficient and mechanical boundary condition.

3.2.2. Experimental data acquisition. According to the welding process parameters in Table1, we input the welding speed, welding current, welding voltage and other data in the welding simulation model, and then input the dynamic thermal physical performance parameters of Q345. After setting the above parameters, we select the layer number components, define the welding line and reference line by node method, calculate the cross-sectional area of the weld bead automatically, determine the parameters of double ellipsoid heat source, calculate the length of the weld bead automatically, and calculate the welding time according to the welding speed, so as to complete the setting of the welding boundary and welding conditions, and output the relationship between thermal-mechanical parameters and temperature. The relationship between thermal-mechanical parameters of Q345 and temperature is shown in Figure 3. At the same time, according to the number of meshes, we automatically set the model into blocks, and set the extraction of welding deformation, welding residual stress and other results [5].

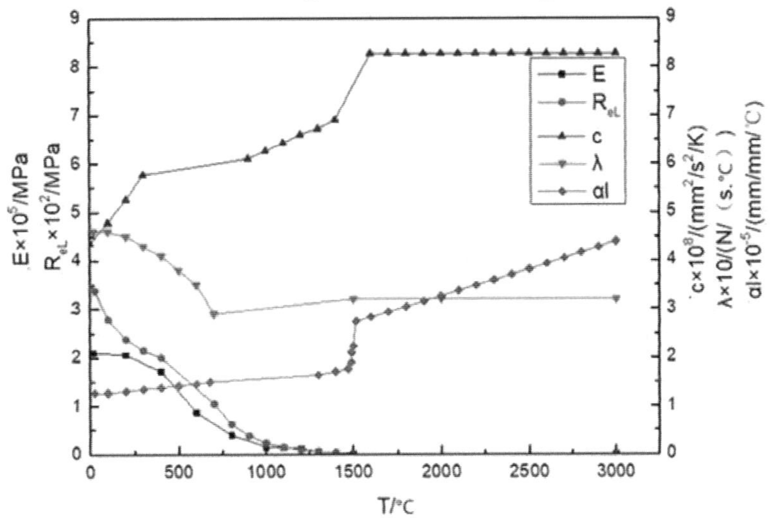

Figure 3. Relationship between thermo-mechanical parameters and temperature of Q345

3.3. Analysis of calculation results.

3.3.1. Welding Deformation Extraction. After the welding simulation calculation is completed, we extract the overall deformation of the butt joint of the experimental panel, as shown in Figure 4. From Figure 4, we can conclude that the main reason for the transverse shrinkage of butt plate is that the base metal expands first in the welding process. When the weld metal solidifies, the expanded base metal will shrink inevitably. This shrinkage is the main component of the transverse shrinkage of butt joint. At the same time, we also find that longitudinal shrinkage occurs along the weld direction, mainly at the head and tail of the weld. The flat plate shrinks in the vertical direction of the weld, and the edge of base metal far from both sides of the weld warps upward, resulting in angular deformation. The deformation trend is consistent with the actual situation [6].

Figure 4. Contour of welding deformation

3.3.2. *Extraction of Welding Residual Stress.* After the calculations, the distribution of welding residual stress of butt joint is extracted, as it can be seen from Figure 5 (a) that with the decrease of the distance from the center of the weld, the transverse residual stress in the vertical direction of the weld increases, reaching the maximum stress in the heat affected zone, the stress on the weld falls back, and the stress in the base metal on both sides of the weld presents a symmetrical distribution; Figure 5 (b) is a nephogram of the longitudinal residual stress along the weld direction. The weld seam and its adjacent area are under tension, and the numerical value generally reaches the yield strength, and the two sides are under compression. The tensile stress in the middle of the weld bead is the largest and decreases gradually toward both ends. After the calculations, the distribution of welding residual stress of butt joint is extracted, as it can be seen from Figure 5 (a) that with the decrease of the distance from the center of the weld, the transverse residual stress in the vertical direction of the weld increases, reaching the maximum stress in the heat affected zone, the stress on the weld falls back, and the stress in the base metal on both sides of the weld presents a symmetrical distribution; Figure 5 (b) is a nephogram of the longitudinal residual stress along the weld direction. The weld seam and its adjacent area are under tension, and the numerical value generally reaches the yield strength, and the two sides are under compression. The tensile stress in the middle of the weld bead is the largest and decreases gradually toward both ends.

IOP Conf. Series: Materials Science and Engineering **637** (2019) 012012 doi:10.1088/1757-899X/637/1/012012

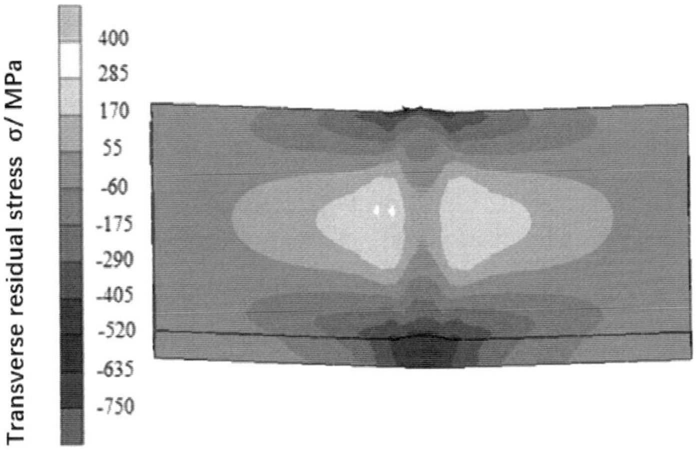

(a) Transverse

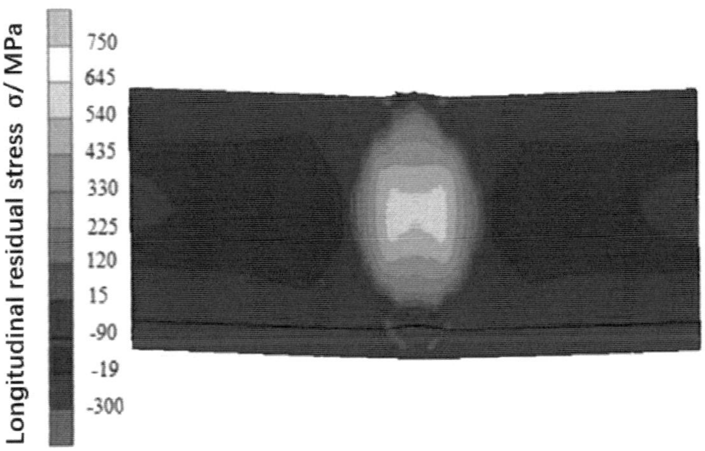

(b) Longitudinal

Figure 5. Contour of welding residual stress distribution

4. Conclusion

In this paper, we simulated the MAG welding process of Q345 steel flat butt weld by using TCL/TK language to build geometric model, mesh generation and model setting of welding simulation software. Through the above research, we can draw a conclusion that the multi-layer butt welding seam of flat plate and its adjacent area are under tension, the numerical value generally reaches the yield strength, and the two sides are under compression. The tensile stress in the middle of the weld bead is the largest and decreases gradually toward both ends. The cross-sectional area of single-sided multi-layer welding seam is larger and asymmetric along the thickness of the plate. The transverse shrinkage and angular deformation of the welded parts are larger after welding. The accuracy of the simulation results is verified by actual welding and testing.

5. Reference

IOP Conf. Series: Materials Science and Engineering **637** (2019) 012012 doi:10.1088/1757-899X/637/1/012012

[1] Jiang Xuewu, Wu Xinyue, Zhu Shijian. Comprehensive application of UG, Hyper Mesh and MSC Marc software for finite element analysis [J]. Computer Aided Engineering, 2007, 16(2):11-14.
[2] Cai Zhipeng, Zhao Haiyan, Wu Ge, et al. Series heat source model and its application in welding numerical simulation [J]. Journal of Mechanical Engineering. 2001, 37 (4): 25-28.
[3] Wei Qiye, Huabin. Dynamic simulation of distillation column under software integration environment [J]. Computer simulation, 2004, 21 (9): 67-70.
[4] Dean Deng, Yijun Zhou, Tao Bi, Xiaozhan Liu. Experimental and numerical investigations of welding distortion induced by CO2 gas arc welding in thin-plate bead-on joints [J]. Materials and Design, 2013 (52): 720-729.
[5] Wang Yudong, Jinlei, Hong Qingquan. HyperMesh & HyperView Application Skills and Advanced Examples [M]. Beijing: Machinery Industry Press, 2012.
[6] Zhang Liping, Fang Yuanbin, Wu Bin, etc. The influence of element type on the accuracy of numerical calculation of welding [J]. Welding Machine, 2016, 46 (11): 88-9

IOP Conf. Series: Materials Science and Engineering **637** (2019) 012013 doi:10.1088/1757-899X/637/1/012013

Finite Element Analysis of Residual Stress in Welded Joints of T91 Steel

Zhi-ling Wang

Jiangsu Province Xuzhou Technician Institute, Xuzhou 221151, China
Email: 379713268@qq.com

Abstract. The pressure equipment is the key equipment in the petrochemical industry, natural gas chemical industry and coal chemical industry. Without safe and reliable pressure equipment, advanced process industry cannot be realized. It is important of the quality assurance of pressure vessel in order to prevent the failure of pressure vessel, especially some rupture accidents. One of the key reasons for quality assurance of pressure vessel is the treatment of residual stress in welded joint. In this paper, the working stress and welding residual stress of the thick plate of pressure equipment material and the welding joint of pressure equipment have been measured and analyzed successively by means of actual stress measurement and finite element simulation analysis. The influencing factors of welding stress are introduced, and the control of residual stress is analyzed and some suggestions are put forward in this paper.

1. Introduction

At present, the main strategic position in the national economy is energy, and boilers are widely used in all walks of life as the most commonly used energy storage containers. The T91 steel is commonly used in supercritical boiler heating surface tubes. SA213T91 is a kind of material between 12Cr1MoV and stainless steel. Its structure is improved 9Cr1Mo high strength martensitic heat resistant steel with high fracture strength and heat corrosion resistance. SA213T91 is an ideal tube for boiler heating surface tubes in the temperature range of 600-650°C. It has good oxidation resistance, higher thermal conductivity than austenitic steel, similar linear expansion coefficient to pearlitic steel, and lower price than austenitic steel [1]. The allowable stress of the steel is higher than that of austenitic stainless steel when the service temperature is lower than 620 °C, especially the dissimilar steel welded joints with austenitic stainless steel. The thermal strength performance of the welded joints can reach the same level as that of the steel itself. It is well known that all high-temperature equipment and pipeline manufacturing is usually completed by welding process, and the use of welding components at high temperature becomes one of the key structural components. In order to ensure the structural integrity and safe and reliable service performance of high temperature components, it is necessary to ensure the reliability of welded joints of high temperature components. In a large number of equipment and pipeline failure accidents, the early failure of welds is often reported. As a weak link of high temperature device system, welded joints of high temperature components have been paid more and more attention by engineering circles. Because of the high concentration of instantaneous heat input in the welding process, there will be huge welding residual stress and deformation in the welding process and after welding.

Through analysis, we can easily conclude that the main reason affecting the safe operation of these components is the residual stress. People have been paying more attention to accurately predicting welding residual stress.

Since the 1980s, the finite element technology has been widely developed. Tekriwal and other scientists have used the finite element method to simulate the arc welding process and predict the welding temperature field [2]. Brust and other scientists used the THERMOELASTIC-PLASTIC finite element model to calculate the welding residual stress of 304 stainless steel pipes for boilers. Professor Tekriwa used three-dimensional finite element transient heat transfer model to simulate the thermal flow in tungsten argon arc welding (TIG) process. Considering the welding speed and current input strength, the width of heat affected zone and weld fusion zone was predicted and compared with the experimental data. Dong has carried out numerical simulation and analysis of two-dimensional and three-dimensional butt welding of circular pipes, and satisfactory results have been obtained [3].

2. Residual Stress Measurement of Welded Joints

2.1. Generation Principle of Welding Stress.

2.1.1 Welding temperature field. The welding process of the boiler is a very uneven local heating process, which will cause uneven shrinkage and expansion of the metal. Therefore, in principle, the stress and deformation caused by welding are the same as that caused by uneven temperature field. However, the stress and deformation of welding are relatively complex, which complexity is shown in the following three aspects： There is a wide range of temperature changes during welding. In the central part of the weld, the highest temperature can reach the boiling point of the material, but the temperature drops sharply when leaving the heat source, so the temperature gradient of the difference is very large; Because the temperature of heating is very high and the range of variation is large, the mechanical properties and thermophysical properties of the metal material vary with the temperature; Because the temperature range of welding varies greatly, so the situation is great. Local phase transition occurs when the phase transition temperature exceeds the phase transition temperature.

2.1.2 Formation of Welding Stress and Welding Deformation. Phase transition will cause many changes in mechanical parameters and physics. For the metals constituting the boiler, after welding, deformation and residual stress will occur in the components, which reason is that the welding area will produce a lot of uneven deformation because of the huge temperature difference, which in turn will be restrained by the effect of deformation around the development of cold metal cannot be free. In the heating stage, because it is not free expansion, it must also emphasize the instantaneous thermal stress on both sides and the tensile stress far away from the weld in the welding and adjacent areas. Because the material has lost its elasticity when the weld and adjacent area T > 800°C, it is limited to the plastic deformation of the expansion and compression of the generated part. In the cooling stage, because it is not free of cooling shrinkage, there will be residual stress [4]. The welding seam and adjacent areas are limited by the shrinkage of residual tensile stress, and the stress is far away from both sides of the weld.

2.2. Preparation of Welded Joints.

The specimens are SA213-T91 straight pipe welding specimens with a diameter of 51mm 8mm. The inert gas shielded welding (TIG) method is used to test the residual stress of pipeline welding and the testing equipment, as shown in Fig1.Heat treatment is required after the completion of welding work. The standard of post-weld heat treatment is as follows. The heating rate is less than 220/h from 420 °C, the holding temperature is 730-760 °C, the holding time is 2 h, the cooling rate after holding is less than 275/h °C, and the cooling temperature is less than 420/h, then it is cooled in static air. The Welding Test Plates and Welding Stress Testing equipment are shown as the figure 1.

IOP Conf. Series: Materials Science and Engineering **637** (2019) 012013 doi:10.1088/1757-899X/637/1/012013

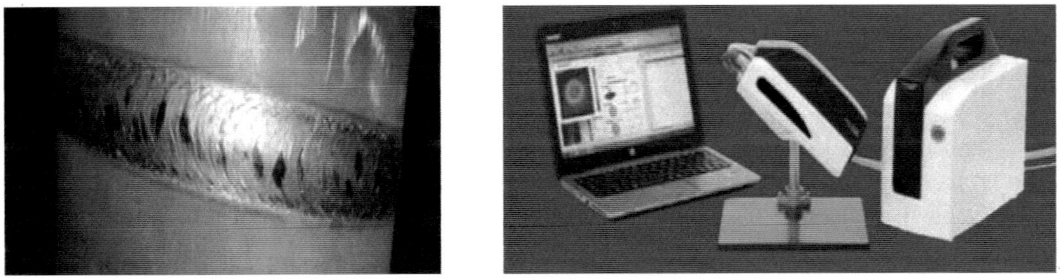

(a) Welded test plate (b) Welding Stress Testing Equipment

Figure 1. Welding Test Plate and Welding Stress Testing Equipment

2.3. Welding Residual Stress Testing.

In order to ensure the accuracy of the test, it is necessary to treat the surface of the test site of the pipeline before X-ray spot strikes the surface of the sample without scattering. The weld surface of this sample has a high smoothness and smoothness, so in order to avoid the pressure stress on the sample surface during grinding with grinding wheel or abrasive paper, this step is cancelled and the sample is electropolished directly until there is enough light and no scratches at the test point. First, we use X-ray diffractometer to measure the axial residual stress on the outer wall of T91 pipeline. Then we cut the pipeline by wire cutting. Then we cut the pipeline along the middle line to measure the axial residual stress on the inner wall. The distribution of welding residual stress test points in the inner and outer walls is shown in Figure 2.

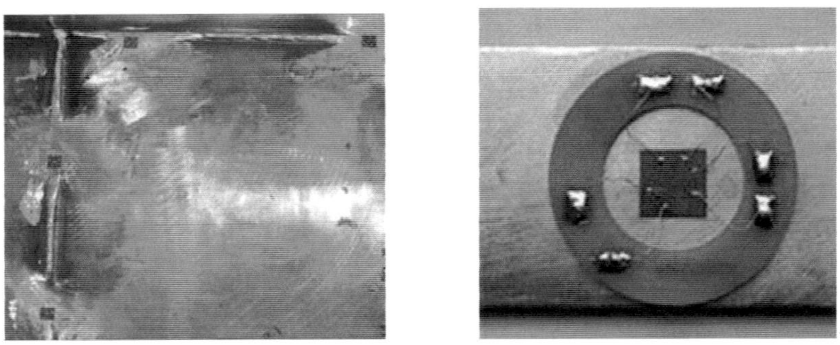

(a)Inside (b) External

Figure 2. Residual Welding Stress of Pipeline Samples

2.4. Comparison of Residual Stress Test Results with Finite Element Results.

In order to verify the validity and reliability of the finite element method for simulating welding residual stress, the test results of axial welding residual stress of inner and outer walls after testing are collated and compared with the results of finite element simulation. By comparing the welding residual stress after heat treatment with that of finite element method, it can be seen that the measured residual stress at the measuring point is close to the corresponding results of finite element simulation, and the distribution trend is also consistent, which can prove that it is feasible to use finite element method to analysis welding residual stress. From the experimental and finite element values, it can be seen that the residual stress at the weld is relatively high, the inner wall is tensile stress, while the outer wall is compressive stress, and the stress value decreases gradually with the increase of the weld distance.

3. Finite Element Analysis of Residual Stress in Welded Joints

3.1. Establishment of Finite Element Model .

The whole finite element analysis process of welded joints is based on the solving process of mesh elements and nodes. In the finite element analysis of welded structure, because of the large heat input at the weld seam and the instantaneous heat conduction, the thermal gradient and stress-strain gradient are the key parts of finite element analysis. The non-linearity of structure and solution equation is very high, which is the key area of finite element analysis [5]. Therefore, the mesh density of the weld and its heat-affected zone increases relatively, while in most base metal areas far from this area, the mesh element size can be gradually enlarged to improve the calculation efficiency. Dimensions and mesh models of T91 steel welded joints are shown in Figure 3.

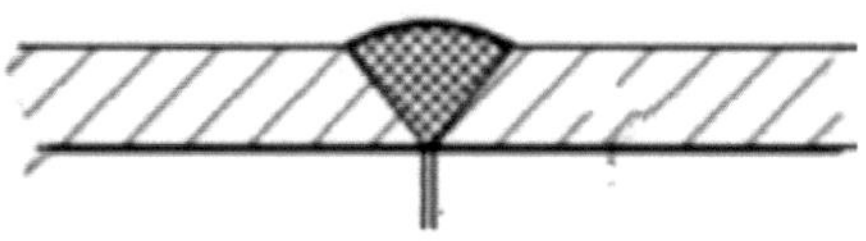

(a)Welded Joint Model

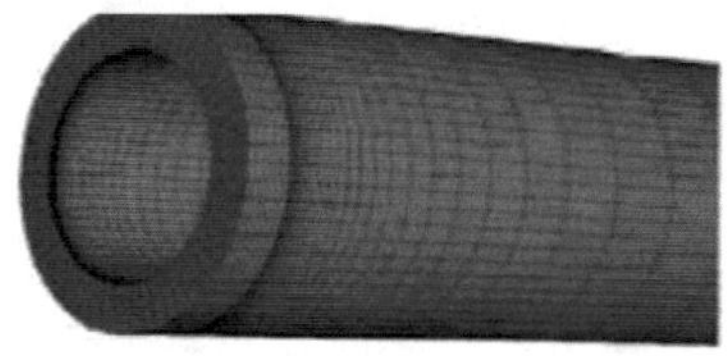

(b) Finite Element Mesh Model

Figure 3. Establishment of Finite Element Model

3.2. Load and boundary conditions.

The load and boundary conditions of the finite element model should simulate the actual welding process as much as possible. For the heat source model of endogenous heat system, the heat source model of thermal welding is used to load the amplitude curve (* AMPLITUDE) into the welding method to simulate the movement of welding heat source by increasing the time. Welding heat source can be effectively converted into heat input per unit time per unit volume of the weld by assuming that internal heating is applied to the weld element. Heat input refers to the heat input from welding energy to the weld of unit length during fusion welding. In the welding process, in order to prevent the whole structure from rigid movement of the welded parts, it is necessary to keep the welded structure away from the weld seam with fixed constraints.

$$q = \eta \mathrm{UI/v}$$

In the formula:
Q - heat input (J/mm);
U - arc voltage (V);
I - welding current (A);
V - Welding speed (mm/s);
η-thermal efficiency (arc welding = 0.7-0.8; submerged arc welding = 0.8-0.95; TIG welding= 0.5).

3.3. Analysis of Welding Stress Simulation Results.

3.3.1. Welding Stress Analysis. Figure 4 shows the residual stresses of the welded joints of straight pipes in the axial and annular welding of the inner wall after as-welded and heat-treated. It can be seen from the diagram that the maximum axial stress and the maximum circumferential stress of the inner wall after welding are 531.9 MPa and 435.4 MPa. As shown in Figure (a), (c), the axial stresses of the heat affected zone and the inner wall weld are tensile stresses. After heat treatment, the maximum axial stress and circumferential stress of the welded joint decreased to 187.9 MPa and 160.8 MPa respectively. After welding heat treatment, the residual stresses were redistributed, and the axial stresses in the inner weld and heat affected zone were more uniform than those in the welded state, and showed tensile stresses. After welding heat treatment, the maximum circumferential stress of the inner wall is still tensile stress.

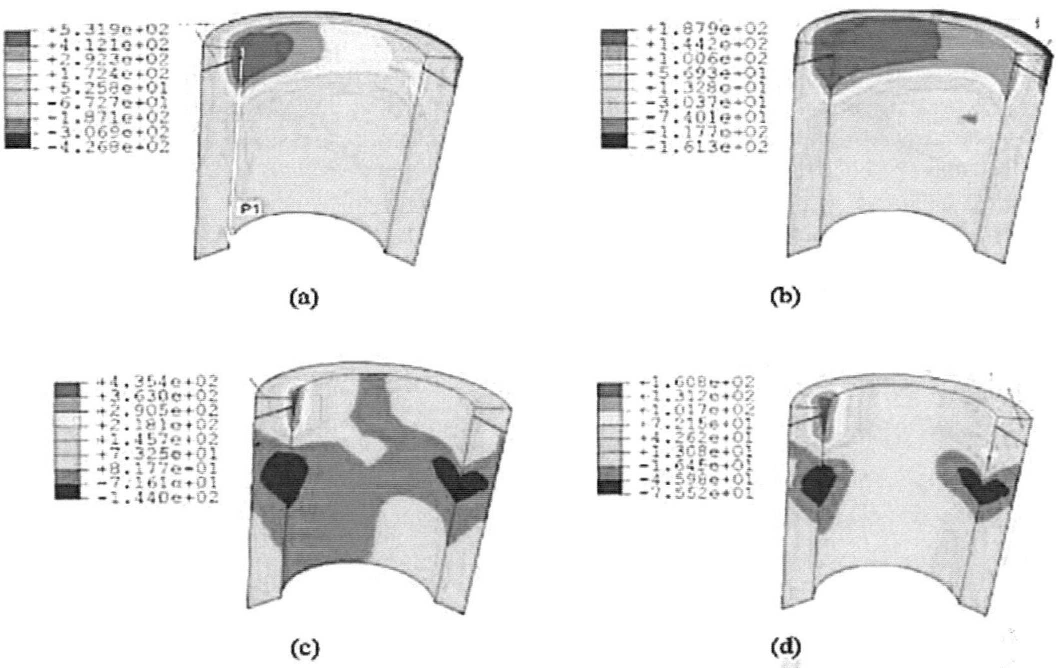

(a) Axial welding stress of welded inner wall; (b) Axis of inner wall after heat treatment; (c) circumferential welding stress of welded inner wall; (d) Circumferential welding stress of inner wall after heat treatment

Figure 4. Distribution of residual stress in axial and circumferential welding of inner wall after as-welded and heat-treated

3.3.2. Postweld heat treatment. The axial and circumferential stresses in the weld seam and heat affected zone of the welded outer wall are compressive stresses, while the axial and circumferential stresses in the weld seam and heat affected zone are compressive stresses. The maximum external axial compressive stress is 426.8 MPA and the maximum circumferential stress is 144 MPA. The pressure level is significantly reduced by post-weld heat treatment, but the pressure gradient is also significantly reduced after stress redistribution. However, the maximum external axial and circumferential stresses are still in the weld and heat affected zone. Maximum axial compressive stress 161.3 MPA, maximum circumferential stress 75.5 MPA. Because post-weld heat treatment can reduce the welding residual stress in the more obvious weld seam and heat-affected zone, the redistribution of post-weld stress can reduce the pressure of residual stress gradient. To learn more about the stress distribution of welded joints and the heat-affected zone of welds perpendicular to the inner and outer reference paths P1 and P2, high residual stress areas can be obtained in the heat-affected zone or in the welded zone. Moreover, the axial and circumferential residual stresses in the weld seam and heat-affected zone are significantly reduced by post-weld heat treatment. High tensile residual stress at welded joints often leads to failure of welded joints under harsh conditions (high temperature, high pressure, corrosion, etc.), such as mode IV cracks and stress corrosion cracks in heat affected zone. At high temperature, the creep strain of welded joints will also be affected by the residual stress when there is a higher residual stress, and the creep strain at the high stress position will increase.

3.4. Equivalent Stress Distribution of Welded Joints.

The experimental results show that when the creep damage process is dominated by the structure coarsening mechanism and there are few voids, it is appropriate to take Von-Mises stress as the equivalent stress of failure under multiaxial stress. The maximum equivalent stress of welded joints occurs in the weld seam and heat affected zone, which is also due to the maximum axial and circumferential equivalent stress at the weld seam. The Mises stress distribution in the outer weld of straight pipe is relatively uniform, ranging from 145 MPa to 188 MPa, while the Mises stress in the local position of inner wall is relatively low, about 50 MPa. If the creep damage process is dominated by microcracks and voids, the failure is controlled by the maximum principal stress. But in most cases, this stress and the maximum principal stress contribute to the failure of high temperature components. According to the fracture stress theory of bone points, the maximum principal stress is a component of equivalent fracture stress that causes failure of high temperature components, which is a good example. The maximum principal stress of the joint is 192.4 MPa at the weld and heat affected zone of the inner wall, while the principal stress of the outer wall is low, almost zero.

4. Conclusion

Through the research, we have a better understanding of the stress in the welded joint of the pressure equipment, and have a thorough analysis of the effective ways to eliminate the stress in the pressure vessel.

4.1. The axial and circumferential stresses of the weld seam and thermal effect of T91 steel are the highest.

In as-welded condition, the stress gradient of weld and base metal is higher, the axial and circumferential stress of inner wall is tensile stress, and the axial and circumferential stress of outer wall is compressive stress. After post-weld heat treatment, the welding residual stress level of the welded joint is obviously reduced, the maximum reduction range is 60%. After heat treatment, the welding residual stress level is reduced to less than 190 MPa, and the maximum reduction is about 50 MPa.

4.2. The maximum principal stress is one of the main reasons to control the failure of high temperature welded structures.

Through the analysis of the maximum principal stress of the component, the equivalent stress level of the weld and heat affected zone is still higher than that of the base metal, which is one of the reasons for the first failure of the welded joint.

4.3. Significance of residual stress analysis of welded joints.

By analyzing the residual stresses of these welded joints, it can be regarded as one of the reasons for the complex loads that high temperature components bear during service, which lays a mechanical foundation for creep damage analysis and life prediction of welded joints.

5. Reference

[1] Wang Jianhua. Welding numerical simulation technology and its application [M]. Shanghai: Shanghai Jiaotong University Press, 2009.
[2] Cai Zhipeng. Research and application of numerical simulation of welding deformation for large structures [D]. Ph.D. Dissertation, Beijing: Tsinghua University, 2001.
[3] Chen Xiaodong. Solutions for eliminating welding residual stress in pressure vessels [J]. Material Engineering, 2003 (6): 15.
[4] Zhang Guodong, Xue Jilin, Zhou Changyu. Welding process optimization of high temperature pipeline based on orthogonal experimental design. Journal of Welding, 2008, 29 (11): 53-56.
[5] Lu An-li, Shi Qingyu, Zhao Haiyan, et al. Several key technical problems in the field of welding process simulation and their preliminary study [J]. China Mechanical Engineering, 2000, 11 (1-2): 201-205.